Marine Electrical Equipment and Practice

Marine Electrical Equipment and Practice

Second edition

H. D. McGeorge, CEng, FIMarE, MRINA

OXFORD AMSTERDAM BOSTON LONDON NEW YORK PARIS
SAN DIEGO SAN FRANCISCO SINGAPORE SYDNEY TOKYO

Butterworth-Heinemann
An imprint of Elsevier Science
Linacre House, Jordan Hill, Oxford OX2 8DP
200 Wheeler Road, Burlington, MA 01803

First published by Stanford Maritime Ltd 1986
Second edition 1993
Reprinted 1995, 1997, 1999 (twice), 2000, 2001, 2003
Transferred to digital printing 2004
Copyright © 1986, 1993, H. David McGeorge. All rights reserved

British Library Cataloguing in Publication Data
McGeorge, H.D.
    Marine Electrical Equipment and
    Practice. – 2Rev.ed
    I. Title
    623.8503

ISBN   0 7506 1647 4

For information on all Butterworth-Heinemann publications
visit our website at www.bh.com

Contents

Preface

The object of this book is to provide a description of the various items of ships' electrical equipment, with an explanation of their operating principles.

The topics dealt with are those that feature in examination papers for Class 1 and Class 2 Department of Transport certification. It is hoped that candidates sitting the electrical paper or the general engineering knowledge paper will find the book helpful in preparing for their examinations.

This second edition includes new chapters on shaft-driven generators and electric propulsion, including many new diagrams explaining drive, distribution and control systems. The treatment of safe electrical equipment has been expanded, and the opportunity has been taken to include reference to more specialised published papers on some of the topics discussed.

Technical language can be a barrier to the understanding of a subject by the non-specialist: an effort has been made to avoid its excessive use, but to explain terms as they arise. Diagrams are kept as simple as possible so that they can form the basis of examination sketches. For this reason many diagrams have been redrawn.

I am grateful for information received from a number of manufacturers of electrical equipment. These include Alcad Ltd; Varta Ltd; NIE-APE-W.H. Allen; Laurence, Scott and Electromotors Ltd; GEC-Alsthom; Clarke, Chapman & Co. Ltd; British Brown Boveri Ltd; and Siemens (UK) Ltd. Much information has been obtained from the Transactions of the Institute of Marine Engineers. Thanks to former colleagues R.C. Dean and R.E. Lovell for their assistance, and also to Alison Murphy for her help with the diagrams.

H.D. McGeorge, CEng, FIMarE, MRINA

CHAPTER ONE

Batteries and Emergency Systems

Lead–acid storage batteries

Each cell of a lead–acid battery contains two interleaved sets of plates, immersed in electrolyte. Those connected to the positive terminal of a charged cell are of lead peroxide; those connected to the negative terminal are of lead. The simple sketch used here to explain the discharge and recharge has only one plate of each type (Figure 1.1).

The electrolyte in which the plates are immersed is a dilute solution of sulphuric acid in distilled water. A characteristic of electrolytes is that they contain ions of the compounds dissolved in them which can act as current carriers. In this solution, the ions are provided by sulphur acid (H_2SO_4) molecules, which split into positively charged hydrogen ions (H^+) and negatively charged sulphate ions (SO_4^{--}). The separated parts of the molecule are electrically unbalanced because the split leaves sulphate ions with extra (negative) electrons, and hydrogen ions with an overall positive charge due to the loss of electrons.

Discharge action

During discharge, the hydrogen ions (H^+) remove oxygen from the lead peroxide (PbO_2) of the positive plates and combine with it to form water (H_2O). Loss of oxygen from the lead peroxide reduces it to grey lead (Pb). The water formed by the action dilutes the electrolyte so that as the cell discharges, the specific gravity (relative density) decreases. Measurement of the specific gravity change with a hydrometer will show the state of charge of the cell.

At the negative side of the cell, sulphate ions (SO_4^{--}) combine with the pure lead of the negative plates to form a layer of white lead sulphate ($PbSO_4$). The lead sulphate layer increases during discharge and finally covers the active material of the plate so that further reaction is stifled. Some sulphate also forms on the positive plates, but this is not a direct part of the discharge reaction.

A fully charged cell will be capable of producing 1.95 volts on load and the relative density of the electrolyte will be at a maximum (say 1.280). After a period of discharge the electrolyte will be weak due to formation of water and the plates will be sulphated, with the result that the voltage on load will drop. Recharging is required when voltage on load drops to say 1.8 volts per cell and the relative density is reduced to about 1.120.

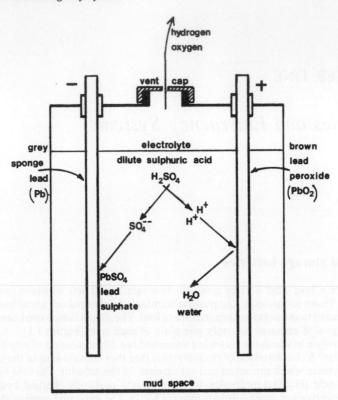

Figure 1.1 *Lead–acid cell*

Charging

To charge lead–acid batteries, the cell is disconnected from the load and connected to a d.c. charging supply of the correct voltage. The positive of the charging supply is connected to the positive side of the cell, and the negative of the charging supply to the negative of the cell. Flow of current from the charging source reverses the discharge action of the cell: thus lead sulphate on the plates is broken down. The sulphate goes back into solution as sulphate ions (SO_4^{--}), leaving the plates as pure lead. Water in the electrolyte breaks down returning hydrogen ions (H^+) to the solution, and allows the oxygen to recombine with the lead of the positive plate and form lead peroxide (PbO_2).

Gas emission

Towards the end of charging and during overcharge, the current flowing into the cell causes a breakdown or electrolysis of water in the electrolyte, shown by bubbles at the surface. Both hydrogen and oxygen are evolved and released through cell vent caps into the battery compartment. There is an explosion risk if hydrogen is allowed to accumulate (flammable range is 4% to 74% of hydrogen in air). Thus regulations require good ventilation to remove gas and precautions against naked lights or sparks in enclosed battery compartments (see below).

Topping up

Batteries suffer water loss due to both gassing and evaporation, with consequent drop in liquid level. There is no loss of sulphuric acid from the electrolyte (unless through spillage). Regular checks are made to ensure that liquid level is above the top of the plates and distilled water is added as necessary. Overfilling will cause the electrolyte to bubble out of the vent.

Plate construction

A lead–acid battery is made up of a number of cells, each with a nominal voltage of 2 volts. Thus three cells separated by divisions in a common casing and connected in series make up a 6 volt battery, and six cells arranged in the same way make a 12 volt battery.

Each cell has, say, seven positive and eight negative plates which are interleaved and arranged alternately positive and negative. Common practice is to have both end plates negative. Plates are prevented from touching by porous separators of insulating plastic. The design of the plates is such as to give the greatest possible surface area, adequate strength and good conductivity. The porous paste active material extends plate area to give maximum contact between active material and the electrolyte, and therefore good capacity. The oxide paste has little strength and is a poor conductor of electricity so the deficiencies are made good by a lead–antimony grid into which the paste is pressed.

Electrolyte

Sulphuric acid used to make up electrolyte for lead–acid batteries is, in its concentrated form, a non-conductor of electricity. In solution with water it becomes an electrolyte because of the breakdown of the H_2SO_4 molecules into hydrogen (H^+) ions and sulphate (SO^{--}) ions which act as current carriers in the liquid.

Concentrated sulphuric acid has a great affinity for water and this, together with the heat evolved when they come into contact, makes the production of electrolyte hazardous. A violent reaction results if water is added to concentrated sulphuric acid. **Successful safe mixing is only possible if the acid is very slowly added to pure water while stirring.** Normally the electrolyte is supplied ready for use in an acid-resistant container.

Electrolyte is strongly corrosive and will damage the skin as well as materials such as paint, wood, cloth etc. on which it may spill. It is recommended that electrolyte on the skin be removed by washing thoroughly (for 15 minutes) with water. Acid-resisting paint must be used on battery room decks.

Nickel–cadmium storage batteries

The active materials of positive and negative plates in each cell of a charged nickel–cadmium battery (Figure 1.2) are nickel hydrate and cadmium, respectively. The chemicals are retained in the supporting structure of perforated metal plates and the design is such as to give maximum contact between active compounds and the electrolyte.

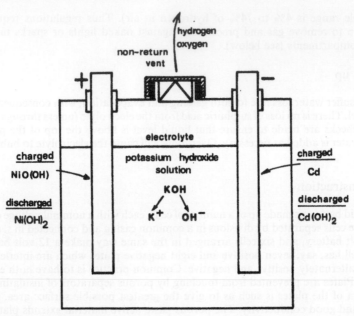

Figure 1.2 *Nickel–cadmium cell*

The strong alkaline electrolyte is a solution of potassium hydroxide in distilled water (with an addition of lithium). The ions produced in the formation of the potassium hydroxide solution (K^+ and OH^-) act as current carriers and take part in an ion transfer.

Discharge action

During discharge the complicated but uncertain action at the positive plates (hydrated oxide of nickel) causes hydroxyl ions (OH^-) to be introduced into the electrolyte. As the action progresses, the nickel hydrate is changed to nickel hydroxide. Simultaneously, hydroxyl ions (OH^-) from the electrolyte form cadmium hydroxide with the cadmium of the negative plates. Effectively, the hydroxyl ions (OH^-) move from one set of plates to the other, leaving the electrolyte unchanged. There is no significant change in specific gravity through the discharge/charge cycle and the state of charge cannot be found by using a hydrometer.

Charging

A direct current supply for charging is obtained from a.c. mains, through the transformer and rectifier in the battery charger. The positive of the charging supply is connected to the positive of the cell, and negative to the negative terminal. Flow of current from the charging source reverses the discharge action. The reactions are complicated but can be summarised by the simplified equation:

Charged		Discharged	
$2NiO(OH)$ + Cd		$2Ni(OH)^2$ H^2O + $Cd(OH)^2$	
Hydrated oxide of nickel	Cadmium	Nickel hydroxide	Cadmium hydroxide

Gassing

The gases evolved during charging are oxygen (at the positive plates) and hydrogen (at the negative plates). Rate of production of gas increases in periods of overcharge. When hydrogen in air reaches a proportion of about 4% and up to 74% it constitutes an explosive mixture. Good ventilation of battery compartments is therefore necessary to remove gas. Equipment likely to cause sparking or arcing must not be located or introduced into battery spaces. Vent caps are non-return valves, as shown diagrammatically (Figure 1.2), so that gas is released but contact by the electrolyte with the atmosphere is prevented. The electrolyte readily absorbs carbon dioxide from the atmosphere and deterioration results because of the formation of potassium carbonate. For this reason, cell vent caps must be kept closed.

Topping up

Gassing is a consequence of the breakdown of water in the electrolyte. This, together with a certain amount of evaporation, means that topping up with distilled water will be necessary from time to time. High consumption of distilled water would suggest overcharging.

Electrolyte

Potassium hydroxide solution is strongly alkaline and the physical and chemical properties of potassium hydroxide closely resemble those of caustic soda (sodium hydroxide). It is corrosive, so care is essential when topping up batteries or handling the electrolyte. In the event of skin or eye contact, the remedy is to wash with plenty of clean water (for 15 minutes) to dilute and remove the solution quickly. Speed is vital to prevent burn damage; and water, which is the best flushing agent, must be readily available. Neutralising compounds (usually weak acids) cannot always be located easily, although they should be available in battery compartments.

Specific gravity of electrolyte in a Ni–Cd cell is about 1.210 and this does not change with charge and discharge as in lead–acid cells. However, over a period of time the strength of the solution will gradually drop and renewal is necessary at about a specific gravity of 1.170.

Containers

The electrolyte slowly attacks glass and various other materials. Containers are therefore of welded sheet steel which is then nickel plated, or moulded in high-impact polystyrene. Steel casings are preferred when battieres are subject to shock and vibration. Hardwood crates are used to keep the cells separate from each other and from the support beneath. Separation is necessary because the positive plate assembly is connected to the steel casing.

Plates

The active materials for nickel–cadmium cells are improved by additions of other substances. Positive plates carry a paste made up initially of nickel hydroxide with a small percentage of other hydroxides to improve capacity and 20% graphite for better conductivity. The material is brought to the charged state by passing a current through it, which changes the nickel hydroxide to hydrated nickel oxide, NiO(OH). Performance of cadmium in the negative plates is improved by addition of 25% iron plus small quantities of nickel and graphite.

Active materials may be held in pocket or sintered plates. The former are made up from nickel plated mild steel strip, shaped to form an enclosing pocket. The pockets are interlocked at their crimped edges and held in a frame. Electrolyte reaches the active materials through perforations in the pockets.

Sintered plates are produced by heating (to 900°C) powdered nickel which has been mixed with a gas-forming powder and pressed into a grid or perforated plate. The process forms a plate which is 75% porous. Active materials are introduced into these voids.

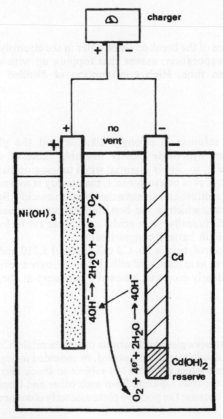

Figure 1.3 *Sealed nickel–cadmium cell*

Sealed nickel–cadmium batteries

Gassing occurs as a conventional battery approaches full charge, and increases during any overcharge due to electrolysis of water in the electrolyte by the current supplied but no longer being used in charging. The gas is released through the vent to prevent pressure build-up and this loss, together with loss from evaporation, makes topping up necessary. While on charge, the active material of the plates is being changed, but when the change is complete and no further convertible material remains, the electrical charging energy starts to break down the electrolyte. Oxygen is evolved at the positive plates and hydrogen at the negative.

Sealed batteries (Figure 1.3) are designed to be maintenance-free and, although developed from and having a similar chemical reaction to the open type, will not lose water through gassing or evaporation. The seal stops loss by evaporation and gassing is inhibited by modification of the plates.

Sealed cells are made with surplus cadmium hydroxide in the negative plate so that it is only partially charged when the positive plate is fully charged. Oxygen is produced by the charging current at the positive plate $(4OH^- \rightarrow 2H_2O + 4e^- + O_2)$ but no hydrogen is generated at the negative plate because some active material remains available for conversion. Further, the oxygen from the positive side is reduced with water at the negative plate $(O_2 + 4e^- + 2H_2O \rightarrow 4OH^-)$, so replacing the hydroxyl ions used in the previous action. The process leaves the electrolyte quantity unaffected. The hydroxyl ions, acting as current carriers within the cell, travel to the positive electrode.

Sealed batteries will accept overcharge at a limited rate indefinitely without pressure rise. Charging equipment is therefore matched for continuous charging at low current, or fast charging is used with automatic cut-out to prevent excessive rise of pressure and temperature. Rise of pressure, temperature and voltage all occur as batteries reach the overcharge area, but the last two are most used as signals to terminate the full charge.

Battery charging

Charging from d.c. mains

The circuit for charging from d.c. mains includes a resistance connected in series, to reduce the current flow from the higher mains voltage. A simple charging circuit is shown in Figure 1.4. Feedback from the battery on charge is prevented, at mains failure, by the relay (which is de-energised) and spring, arranged to automatically disconnect the battery. The contacts are spring operated; gravity opening is not acceptable for marine installations.

Charging from a.c. mains

Mains a.c. voltage is reduced by transformer to a suitable value and then rectified to give a direct current for charging. The supply current may be taken from the 230 volt section and changed to say 30 volts for charging 24 volt batteries. Various transformer/rectifier circuits are described in Chapter 2 and any of these could be used (i.e. a single diode and half-wave rectification, two or four diodes and full-wave rectification, or a three-phase six diode circuit). Smoothing is not essential for battery charging but would be incorporated for power supplies to low-pressure d.c. systems with standby batteries, and for systems with batteries on float.

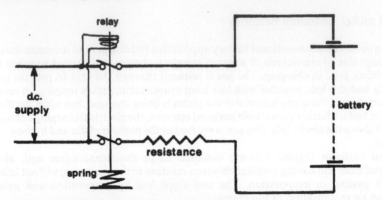

Figure 1.4 *Battery charging from a direct current supply*

The circuit shown (Figure 1.5) has a transformer and bridge of four diodes with a resistance to limit current. The resistance is built into the transformer secondary by many manufacturers. Voltage is dropped in the transformer and then applied to the diodes which act as electrical non-return valves. Each clockwise wave of current will travel to the batteries through D_1 and return through D_2 (being blocked by the other diodes). Each anti-clockwise wave will pass through D_3 and back through D_4. Thus only current in one direction will reach the batteries.

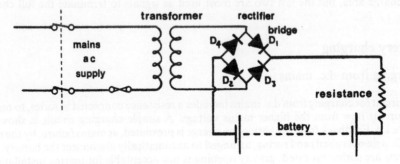

Figure 1.5 *Battery charging from alternating current*

Standby emergency batteries

Emergency power or temporary emergency power can be provided by automatic connection of a battery at loss of main power. A simple arrangement is shown (Figure 1.6) for lead–acid batteries. This type of secondary cell loses charge gradually over a period of time. Rate of loss is kept to a minimum by maintaining the cells in a clean and dry state, but it is necessary to make up the loss of charge: the system shown has a trickle charge.

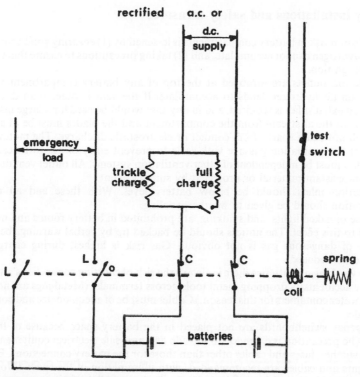

Figure 1.6 *Emergency battery circuit*

In normal circumstances the batteries are on standby with load switches (L) open and charging switches (C) closed. This position of the switches is held by the electromagnetic coil against pressure of the spring. Loss of main power has the effect of de-energising the coil so that the switches are changed by spring pressure moving the operating rod. The batteries are disconnected from the mains as switch C opens, and connected to the emergency load by closing of L.

Loss of charge is made up when the batteries are on standby, through the trickle charge which is adjusted to supply a continuous constant current. This is set so that it only compensates for losses which are not the result of external load. The current value (50 to 100 milliamperes per 100 ampere hours of battery capacity) is arrived at by checking with a trial value that the battery is neither losing charge (hydrometer test) nor being overcharged (gassing).

When batteries have been discharged on load the trickle current, set only to make up leakage, is insufficient to recharge them. Full charge is restored by switching in the quick charge. Afterwards batteries are put back on trickle charge.

Battery installations and safety measures

The explosion risk in battery compartments is lessened by (1) ensuring good ventilation so that the hydrogen cannot accumulate, and (2) taking precautions to ensure that there is no source of ignition.

Ventilation outlets are arranged at the top of any battery compartment where the lighter-than-air hydrogen tends to accumulate. If the vent is other than direct to the outside, an exhaust fan is required, and in any case would be used for a large installation. The fan is in the airstream from the compartment and the blades must be of a material which will not cause sparks from contact or electrostatic discharge. The motor must be outside of the ventilation passage with seals to prevent entry of gas to its casing. The exhaust fan must be independent of other ventilation systems. All outlet vent ducts are of corrosion-resistant material or protected by suitable paint.

Ventilation inlets should be below battery level. With these and all openings, consideration should be given to weatherproofing.

The use of naked lights, and smoking, are prohibited in battery rooms and notices are required to this effect. The notices should be backed up by verbal warnings because the presence of dangerous gas is not obvious. Gas risk is highest during charging or if ventilation is reduced.

When working on batteries there is always the risk of shorting connections and causing an arc by accidentally dropping metal tools across terminals. (Metal jugs are not used as distilled water containers for this reason.) Cables must be of adequate size and connections well made.

Emergency switchboards are not placed in the battery space because of the risk of arcing. The precaution is extended to include any non-safe electrical equipment, battery testers, switches, fuses and cables other than those for the battery connections. Externally fitted lights and cables are recommended, with illumination of the space through glass ports in the sides or deckhead. Alternatively, flameproof light fittings are permitted.

Ideal temperature conditions are in the range from 15 °C to 25 °C. Battery life is shortened by temperature rises above 50 °C, and capacity is reduced by low temperatures.

Emergency generator

There are a number of ways in which emergency power can be supplied. The arrangement shown in Figure 1.7 incorporates some common features.

The emergency switchboard has two sections, one operating at 440 volts and the other at 220 volts. The 440 volt supply, under normal circumstances, is taken from the main engineroom switchboard through a circuit breaker A. Loss of main power causes this breaker to be tripped and the supply is taken over directly by the emergency generator when started, through breaker B. An interlock prevents simultaneous closure of both breakers.

A special feeder is sometimes fitted so that in a dead-ship situation the emergency generator can be connected to the main switchboard. This special condition breaker would only be closed when the engineroom board was cleared of all load, i.e. all distribution breakers were open. Selected machinery within the capacity of the emergency generator could then be operated to restore power, at which stage the special breaker would be disconnected.

The essential services supplied from the 440 volt section of the emergency board depicted include the emergency bilge pump, the sprinkler pump and compressor, one of

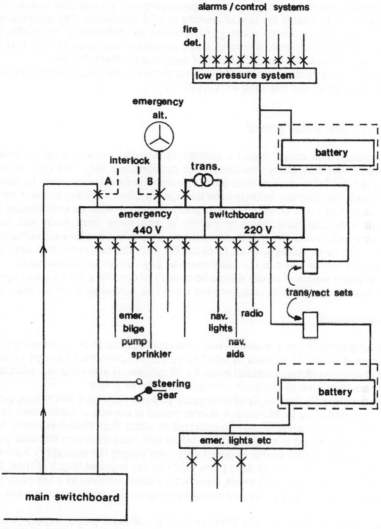

alarms / control systems

fire
det.

low pressure system

emergency
alt.

interlock

trans.

A | | B

emergency   switchboard
440 V   220 V

battery

trans/rect sets

emer.   nav.   radio
bilge   lights
pump   nav.
sprinkler   aids

steering
gear

battery

emer. lights etc

main switchboard

Figure 1.7 *Emergency power supply*

two steering gear circuits (the other being from the main board), and a 440/220 volt three-phase transformer through which the other section is fed.

Circuits supplied from the 220 volt section include those for navigation equipment, radio communication and the transformed and rectified supplies to battery systems. Separate sets of batteries are fitted for temporary emergency power and for a low-pressure d.c. system. The former automatically supply emergency lights and other services not connected to the low-pressure system. Batteries for the radio are not shown.

The switchboard and generator for emergency purposes are installed in one compartment which may be heated for ease of starting in cold conditions. The independent and approved means of automatic starting (compressed air, batteries or hydraulic) should have the capacity for repeated attempts, and a secondary arrangement such that further attempts can be made within the 30 minute temporary battery lifetime.

The emergency generator is provided with an adequate and independent supply of fuel with a flash point of not less than 43 °C (110 °F).

Emergency electrical power

In all passenger and cargo vessels a number of essential services must be able to be maintained under emergency conditions. The requirements vary with type of ship and length of voyage. Self-contained emergency sources of electrical power must be installed in positions such that they are unlikely to be damaged or affected by any incident which has caused the loss of main power. The emergency generator with its switchboard is thus located in a compartment which is outside of and away from main and auxiliary machinery spaces, above the uppermost continuous deck and not forward of the collision bulkhead. The same ruling applies to batteries, with the exception that accumulator batteries must not be fitted in the same space as any emergency switchboard.

An emergency source of power should be capable of operating with a list of up to $22\frac{1}{2}°$ and a trim of up to 10°. The compartment should be accessible from the open deck.

Passenger vessels

Emergency generators for passenger vessels are now required to be automatically started and connected within 45 seconds. A set of automatically connected emergency batteries, capable of carrying certain essential items for 30 minutes, is also required. Alternatively, batteries are permitted as the main emergency source of power.

Regulations specify the supply of emergency power to essential services on passenger ships for a period of up to 36 hours. A shorter period is allowed in vessels such as ferries. Some of the essential services may be operated by other than electrical means (such as hydraulically controlled watertight doors), others may have their own electrical power. If the batteries are the only source of power they must supply the emergency load without recharging or excessive voltage drop (12% limit) for the required length of time. Because the specified period is up to 36 hours, batteries are used normally as a temporary power source with the emergency generator taking over essential supplies when it starts (Figure 1.7).

Batteries are fitted to provide temporary or transitional power supply, emergency lights, navigation lights, watertight door circuits including alarms and indicators, and internal communication systems. In addition they could supply fire detection and alarm installations, manual fire alarms, fire door release gear, internal signals, ship's whistle and daylight signalling lamp. But some of these will have their own power or take it from a low-pressure d.c. system. Sequential watertight door closure by transitional batteries is acceptable.

The emergency generator when started supplies essential services through its own switchboard, including the load taken initially by the transitional batteries. Additionally it would provide power for the emergency bilge pump, fire pump, sprinkler pump, steering gear and other items if they were fed through the emergency switchboard.

Arrangements are required to enable lifts to be brought to deck level in an emergency.

Also, emergency lighting from transitional batteries is required in all alleyways, stairs, exits, boat stations (deck and overside), control stations (bridge, radio room, engine control room etc.), machinery spaces and emergency machinery spaces.

Cargo vessels

Emergency power for cargo ships is provided by accumulator battery or generator. Battery systems are automatically connected upon loss of the main supply, and in installations where the generator is not started and connected within 15 seconds automatically, are required as a transitional power source for at least 30 minutes.

Power available for emergencies must be sufficient to operate certain essential services simultaneously for up to 18 hours. These are: emergency lights, navigation lights, internal communication equipment, daylight signalling lamp, ship's whistle, fire detection and alarm installations, manual fire alarms, other internal emergency signals, the emergency fire pump, steering gear, navigation aids and other equipment. Some essential services have their own power or are supplied from a low-pressure d.c. system.

Transitional batteries are required to supply for 30 minutes power for emergency lighting, general alarm, fire detection and alarm system, communication equipment and navigation lights.

CHAPTER TWO

Electronic Equipment

An explanation of the operation of electronic equipment requires consideration of atomic structure, and the part played by the outer valence electrons in the bonding of atoms and in current carrying.

The structure of an atom

The atom is familiarly depicted by a simple model showing a central nucleus which contains neutrons and positively charged protons, surrounded by orbiting negative electrons (Figure 2.1). The negative charge of each of the electrons is equal but opposite to

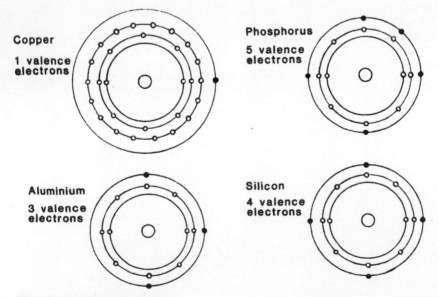

Figure 2.1 *Simple atomic structure*

the positive charge of each of the protons. Electrical balance of the individual atom requires that the number of positive protons in the nucleus is matched by an equal number of orbiting negative electrons. Loss or gain of electrons changes the atom to a positively or negatively charged ion due to the alteration in this balance.

Valence electrons

Electrons orbiting at the greatest distance from the nucleus of an atom are termed valence electrons. They play a major role in the bonding together of atoms in materials or compounds. Valence electrons are also able to detach themselves and take part in the movement of electrons associated with the flow of electrical current through an electrical conductor. In an insulator, the outer electrons are not easily detached.

Conductors and insulators

That metals are good conductors is explained by the large number of electrons which are present in the structure and available to take part in current flow. Insulators have a very stable atomic structure in which orbiting outer electrons are bound tightly to the nucleus. Good insulators (e.g. glass, rubber, various plastics) are all compounds, not elements.

Semi-conductor materials

The semi-conductor materials silicon and germanium are, in the pure state, elements which are neither good conductors like copper nor good insulators such as glass and rubber. Conductivity is related to the number of electrons freely available to take part in current flow, and the valence electrons in the semi-conductor materials are bound into the crystal structure sufficiently tightly to make them poor conductors but not well enough to make them insulators.

Atoms of silicon and germanium (Figure 2.2) have four valence electrons which take part in covalent bonds within the crystal structure. There are alternative ways of depicting the bonding arrangement (Figure 2.3) but only the important valance electrons are shown

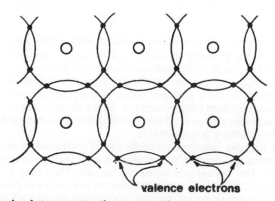

valence electrons

Figure 2.2 *Covalent bonds in a semi-conductor material*

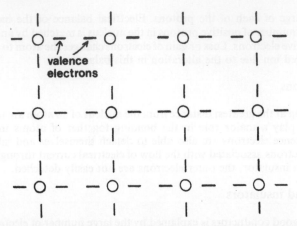

Figure 2.3 *Alternative method of showing covalent bonds in a semi-conductor material*

here. Each atom contributes one electron to the bond with another and also provides a space or **hole** into which an electron from the other atom can fit. Having four electrons for bonding means that each atom is associated with four others in the covalent bonds. At absolute zero temperature all of the valence electrons in a pure crystal of semi-conductor material are firmly held, making the material an insulator. The effect of heat energy from only a moderate temperature is that thermal agitation will cause a few electrons to become detached and able to move about. Each leaves a hole in its system which tends to attract another electron, so leaving another hole etc. The result of heat energy is the random movement of a few freed electrons (negative charges) between holes, and equally, random movement of the positions of the holes. Heat thus makes current carriers available and at ambient temperatures semi-conductors will carry a very small leakage current if a voltage is applied. The number of current carriers released by thermal effect increases with temperature, as does the current flow.

Thermistors

Conductivity of a semi-conductor material increases with temperature rise and falls as temperature drops. (This property is also found in carbon and the characteristic is opposite to that found in metals, whose resistances increase with rise in temperature.) The explanation for the negative temperature coefficient of resistance in silicon and germanium is that valence electrons forming covalent bonds in the structure are displaced due to increase of energy resulting from temperature rise. As the electron wanders it leaves a hole, and in effect two charge carriers are produced by break-up of an electron hole pair. At very low temperature the structure is more stable.

The property of **negative temperature coefficient** of resistance is used in devices for temperature measurement, thermal compensation and thermal relays for protection against overheating. The devices are called *thermistors*. They may be produced from materials other than silicon or germanium because the properties of these are sensitive to impurities. Preferred materials are mixtures of certain pure oxides such as those of nickel, manganese and cobalt, which are sintered with a binding compound to produce a small bead with a negative coefficient of resistance.

Thermistors with the opposite characteristic, of a rising resistance with increase of temperature, are preferred for some applications. They are described as having a **positive coefficient of resistance.**

P- and N-type semi-conductors

Semi-conductor materials such as silicon and germanium, in the pure state and at ambient temperature, are able to pass only a minute leakage current when moderate voltage is applied to them. They act as inferior insulators rather than conductors, because of the lack of current carriers. They can, however, be made to conduct freely by the addition of small traces of certain elements.

N-type semi-conductors

N(for negative)-type materials have enhanced conductivity as the result of the addition of an impurity with five valence electrons instead of the four normal to semi-conductor materials. Antimony, arsenic and phosphorus are elements which could be used to 'donate' an extra free electron.

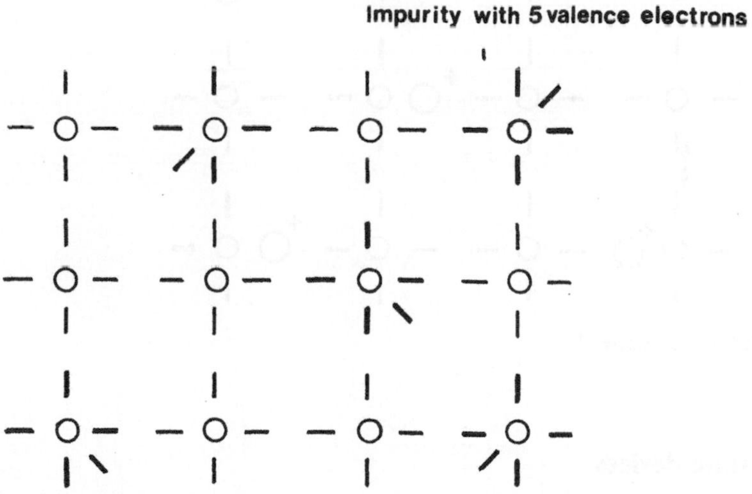

Figure 2.4 *N-type material*

The atoms of the impurity used are scattered through the semi-conductor material (Figure 2.4) with four of their valence electrons joining with the atoms of the parent material in covalent bonds and the fifth left free in the structure to act as a current carrier. Full current flow in an n-type material is based on the movement of these electrons.

P-type semi-conductors

P-type semi-conductors are made by adding impurities such as aluminium, borium or indium. Each of these is able to produce the apparent effect of removing negative electrons, so leaving gaps or positive holes which transmit charge. The p-type impurities are sometimes termed 'acceptors' for this reason.

The explanation is that the additive atoms have only three valence electrons and can only participate in covalent pair bonds with three, not four, of the surrounding semi-conductor atoms (Figure 2.5). The hole available for the fourth valence electron is left vacant, and because it represents the lack of a negative electron is considered the equivalent of a positive charge. Such a positive hole has an attraction for valence electrons of adjacent atoms, but when filled by one of these another hole is left. Random movement of the hole would result from high-energy electrons moving into the gap and leaving another. Current flow through p-type semi-conductor material involves the movement of holes from the positive terminal of the power supply towards the negative terminal.

Impurity with 3 valence electrons

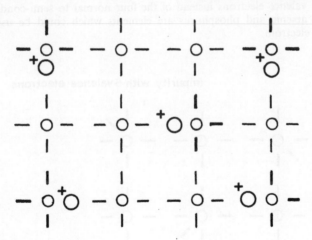

Figure 2.5 *P-type material*

Solid state devices

Various techniques can be used (Figure 2.6) to produce devices having two, three, four or five layers which are alternately p and n, in a single homogeneous crystal or slice of semi-conductor material. Current is transferred through the layers by movement of negative electrons (as in metals) and of positive holes.

The term 'solid state' is used to describe these devices because the current carriers move through the solid material, unlike electrons in thermionic valves which move in space to cross the gap between cathode and anode, or current carrying ions moving between electrodes in a liquid.

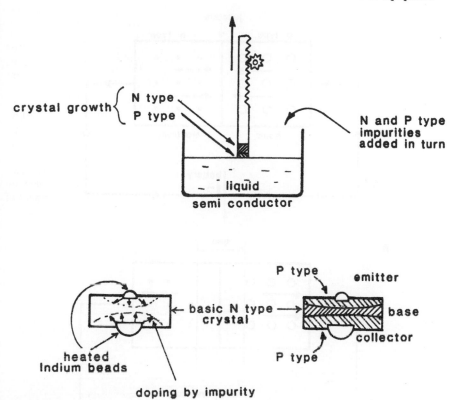

Figure 2.6 *Two methods used to produce pn devices*

Semi-conductor junction rectifier (two-layer device)

A semi-conductor junction rectifier is a wafer of silicon, germanium or other semi-conductor material which has been doped by the impurities mentioned above, or by other materials having similar effects, so that one part is p-type and the other is n-type (Figure 2.7). A battery connected in circuit will cause positive holes in the p-section to be attracted towards its negative terminal. Negative electrons in the n-section tend to move to the positive of the battery. Both positive and negative carriers therefore move towards the junction, when the battery is as shown in Figure 2.7a, and current is conducted in the circuit and across between the two regions. If the battery is connected the other way round the effect is still that positive charges are attracted to the negative, and negative electrons to the positive terminals (Figure 2.7b), but no current flows because both carriers move away from the junction, leaving a gap in the circuit. The ability of the rectifier to switch off when reverse biased enables it to be used as the means of converting alternating to direct current.

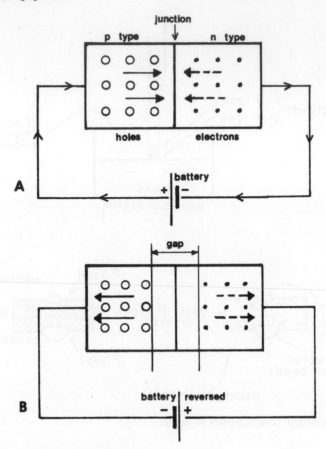

Figure 2.7 *Semi-conductor junction rectifier*

Characteristic curves

Current–voltage characteristics for a semi-conductor junction rectifier are completely different when voltage is applied in the forward direction compared with the effect of reverse voltage (Figure 2.8). This is to be expected in a device that freely conducts when forward biased and resists backward current flow. In order to obtain a reasonable curve, however, the scales are not related on either axis between forward and reverse parts.

A silicon rectifier shows a rise in current flow as forward voltage reaches about 0.5 volts. Very large current flow is produced by quite small further voltage increases.

Reverse voltage causes negligible 'leakage' current up to a value at which reverse breakdown occurs. This breakdown voltage (sometimes called zener or avalanche voltage) varies from one diode to another. The leakage current depends on a few current carriers liberated by ambient temperature and because these are usually small in number, leakage current remains negligibly low and almost constant. Increase of ambient

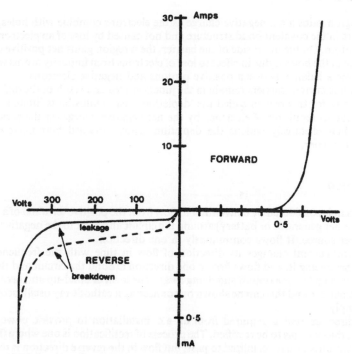

Figure 2.8 *Characteristic curve for a semi-conductor junction rectifier*

temperature will increase leakage current slightly by releasing more electron/hole carriers. The increased leakage from a small temperature rise is shown by the dotted line.

Majority and minority carriers

Majority carriers in p- and n-type semi-conductors are, respectively, holes and electrons, which are present as the result of doping with impurities. Main current flow in p-type materials is associated with hole movement. In n-type materials the main current flow is involved with movement of electrons.

Minority carriers are those present in a doped semi-conductor material as the result of factors other than the doping. Thus leakage current in reverse biased diodes (see Figure 2.8) occurs due to a few minority carriers present as the result of thermal energy breaking up electron/hole pairs.

In other solid state devices electrons may be emitted into p-type sections or holes into n-type sections, when they are termed minority carriers.

Barrier potential

A phenomenon frequently referred to in descriptions of activity at pn junctions is the potential built up by diffusion of negative electrons from n to p regions and the appearance of holes on the n side.

The p region gains a net negative charge as the electrons combine with holes, because the holes are in the covalent bond structure and not caused by loss of an electron from an individual atom. On the other side of the barrier, the n region gains net positive charge as the presence of the holes is due in effect to loss of electrons from impurity atoms which had in themselves a balance between positive protons and negative electrons.

Very few free current carriers remain in the junction area as a result of the diffusion and it is sometimes for this reason called the 'depletion area'. Diffusion is limited as charge builds up due to repulsion of electrons by the net negative charge on the p side.

Reverse bias effectively widens the depletion layer. Forward bias narrows it and removes the barrier.

Rectification

Direct current is considered by convention to flow from the positive terminal of a source of supply (whether generator or battery) around the circuit and back to the negative terminal of the power source. It flows continuously in one direction.

Alternating current changes its direction of flow in time with the frequency of the alternator producing it, and flows first in one direction around the circuit and then in the other, building up to a maximum and dying away. The form of build-up and decay follows a sine wave pattern and this can be shown by connecting a cathode ray oscilloscope across the a.c. supply.

Where direct current is required from an a.c. installation to provide power for d.c. equipment, then a.c. has to be rectified. The process of rectification is one where the flow of current in one direction is permitted to pass, but flow in the reverse direction is resisted, or channelled in the other direction.

The modern rectifying device is a **semi-conductor junction rectifier**. These are pn diodes which, like the thermionic valves and metal rectifiers formerly used, act as electrical non-return valves when connected into an alternating current circuit. That is, they permit current flow in one direction but resist a reverse flow.

Transformers in rectifier circuits

Transformers are included in rectifier circuits for battery systems to bring a.c. mains voltage down to the required level. Voltage of alternating supplies (but not direct current) can be increased or decreased with very small power loss. A transformer also improves the safety of an installation by isolating mains from the equipment being supplied. (See Chapter 4.)

Half-wave rectification

Figure 2.9 shows a transformed a.c. supply connected to a load with a rectifier (or electrical non-return valve) in the circuit.

Referring to the secondary winding of the transformer, when terminal T is positive relative to terminal B, conventional current flows in a direction that agrees with that of the arrow symbol representing the rectifying diode. Current passes through the rectifier to the load and the rectifier is said to be forward biased. When the situation changes and B is positive relative to T, then current flow in the circuit would tend to be the other way. This flow is resisted by the rectifier.

The effect of the single rectifier is to produce half-wave rectification and, as with

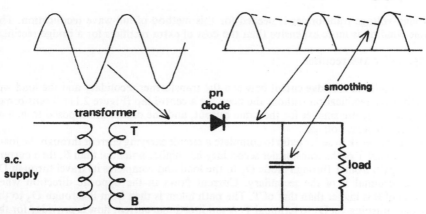

Figure 2.9 *Half-wave rectification*

alternating current, this can be demonstrated using a cathode ray oscilloscope. The half sine waves indicate unidirectional although not continuous flow of current through the load as a result of the pattern of voltage developed. To obtain a d.c. supply with less ripple, the pulsations can be reduced by a capacitor smoothing circuit.

Full-wave rectification

Both half-cycles of the alternating current input can be applied to the load with an arrangement of two diodes and a transformer having a centre tap (Figure 2.10). Each pn diode conducts in turn when the end of the secondary winding which supplies it has full potential relative to the centre tap.

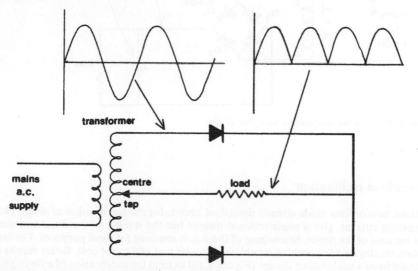

Figure 2.10 *Full-wave rectification with two diodes*

A high-voltage transformer is needed for this method of full-wave rectification. The double winding is more expensive than the cost of extra rectifiers for a bridge rectifier.

Bridge full-wave rectifier

Four pn diodes in a bridge circuit between the transformer secondary and the load will give full-wave rectification without the need for a centre tap (Figure 2.11). Transformer voltage and size are smaller for the same output, and the diodes are exposed to half as much peak reverse voltage.

The diodes work in series pairs to complete a circuit carrying current through the load. When terminal T of the transformer secondary has higher potential than B, then current follows a path from T through diode D_1 to the load and completes its travel through D_2 back to terminal B of the secondary. Current flows in the opposite direction when potential of B is higher than that of T. The path taken is then from B through D_3 to the load and returning via D_4 to terminal T. A unidirectional current flow is provided for the load and smoothing can be applied to reduce ripple.

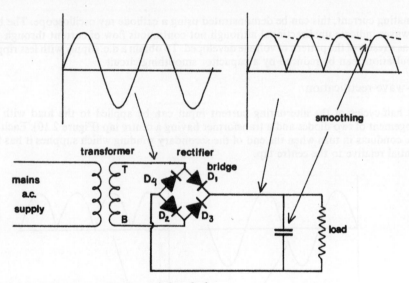

Figure 2.11 *Full-wave rectification with four diodes*

Three-phase rectification

The one, two or four diode circuits described above, for the rectification of single phase alternating current, give a unidirectional output but the quality of the direct current is poor because of the ripple. Smoothing of the d.c. is required for most purposes. For large powers, the degree of smoothing would cause losses and add to the cost. Better results are obtained from a bridge of six diodes (Figure 2.12) as used for rectification of a three-phase supply.

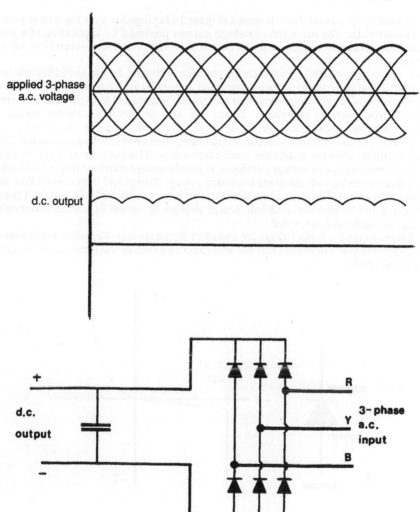

Figure 2.12 *Three-phase rectifier*

Zener diodes (two-layer devices)

A semi-conductor junction rectifier connected in reverse will block current flow unless the voltage is raised to the reverse breakdown level. This feature is shown in the section on these devices, by the characteristic curves. Zener diodes are components which are constructed in the same way but made with a specific reverse breakdown voltage, by controlling the manufacturing process. Breakdown voltage is governed by the material, amount of impurity added and thickness of layers. It can be varied from one to several hundred volts.

The symbol for a zener diode is shown (Figure 2.13a) together with the reverse part of the characteristic. The curve shows leakage current produced by application of a small reverse voltage and then massive current flow as the result of exceeding reverse breakdown voltage.

The breakdown capability makes zeners useful as protective devices and in this role they act as electrical relief valves. A zener connected in parallel with a sensitive meter can, by incorporation of suitable resistances into the circuit, be arranged to bypass excess current resulting from increase of voltage. Zeners are able to perform a similar service in battery-charging circuits and in shunt diode safety barriers for intrinsically safe equipment. The characteristic curve shows that after breakdown, voltage across the diode remains almost constant despite increased current flow. This feature makes zeners useful as voltage references and voltage stabilisers. A simple voltage stabiliser (Figure 2.13b) has a resistance in series with the diode to absorb voltage change and limit current flow as a protective measure. The zener connected so as to be reverse biased will, provided that it operates in the breakdown condition, accept changes in current flow while maintaining voltage at the breakdown value.

Changes to the d.c. input voltage are absorbed by the resistor. The effect on the zener is that its current flow will rise or drop, but zener and load voltage will remain constant at the breakdown point.

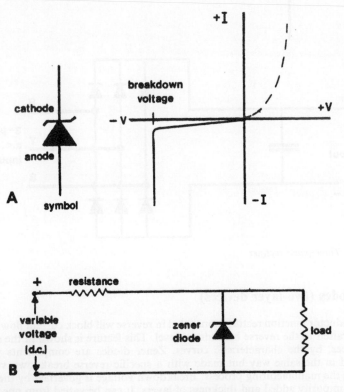

Figure 2.13 *Zener diode characteristic and voltage stabilisation circuit*

If load current increases, the zener current drops by the same amount. Fall in load current causes a rise in zener current. Again the diode acts as a ready bypass for changing current without altering voltage, which remains at the breakdown figure.

Voltage stabilisers are fitted to maintain constant voltage from variable d.c. supplies. An example of their use is in battery systems where voltage drops during the discharge period of the cells. Equipment operates more efficiently from a stabilised supply. Battery voltage must always be above that required by the system and above breakdown voltage of the zener. Sufficient cells must be incorporated for this.

Voltage doubler

The circuit shown (Figure 2.14) can be used to produce a voltage across the load which is almost twice the peak value of the a.c. input, provided that the load current is small. Similar circuits are available to increase voltage by a factor of almost four, but only if load current is very small.

The voltage doubler has two semi-conductor junction rectifiers and two capacitors connected, in this version, in the form of a bridge. Alternating current from the a.c. input will flow clockwise and then anticlockwise, as shown by the arrows. Clockwise flow via rectifier R_1 will charge capacitor C_1 to approximately the peak of the positive voltage wave. Anticlockwise flow, also shown by arrows, through rectifier R_2 will charge capacitor C_2 to approximately the peak of the negative voltage wave. The two capacitors are in series with each other and also with the load. They discharge to the load, and being in series their opposing voltages add to give a doubling effect.

Voltage multipliers are used in television and radar equipment as an alternative to a heavier, larger and more expensive transformer and rectifier arrangement.

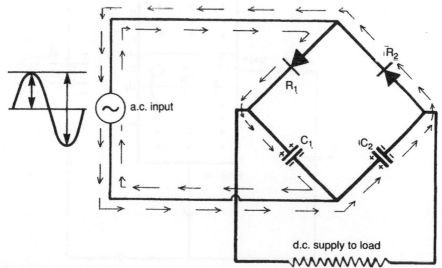

Figure 2.14 *Voltage doubler*

Transistors

Rectifiers and zener diodes are two-layer semi-conductor devices. Transistors have three layers which are arranged as either **npn** or **pnp**. Methods used to make transistors are similar to those for the manufacture of diodes. Their operation is based on the principle that application of voltage will make negative current carriers move in one direction and positive carriers in the other.

Operation of npn transistors

The simple sketch of an npn transistor (Figure 2.15) shows a battery A connected to the ends marked collector and emitter. Another battery B is connected to the middle base section and has a common connection with circuit A to the emitter.

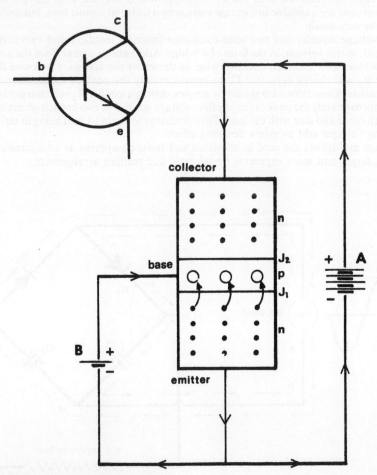

Figure 2.15 *Transistor with npn arrangement*

With only battery A in circuit, no current passes through the transistor because junction J_2 is reverse biased by the battery. That is, negative electrons are attracted to the battery positive terminal and positive holes to the negative. Both sets of current carriers move away from the junction and no interchange occurs to promote current flow. (Junction J_1 is pn forward biased relative to battery A and would pass current.)

When battery B is connected, it has the effect of causing current to flow across junction J_1 because holes in the base are attracted to the negative terminal of B and electrons in the emitter layer are attracted to the positive. The electrons fill the holes and produce an excess of electrons in the p section base. The base is made thin and doped with a small amount of impurity to promote this effect.

Electrons emitted into the base by battery B (minority carriers) act as current carriers for flow from battery A. The base becomes temporarily n-type. Current flow from battery A is governed by the number of carriers injected and this in turn depends on the strength of the voltage signal from B. If the strength of the small input signal from B is varied, the result will be a change in current flowing from battery A and through the transistor. Use of a transistor to control a large current with a small input signal is called amplification.

The symbol above Figure 2.15 for an npn transistor has an arrow on the emitter connection pointing away from the vertical line representing the base. The emitter, base and collector connections are also denoted by their initial letters but these are unnecessary. The base forms a T with the middle base connection; collector and emitter are at either side. The arrow points in the direction of conventional current flow.

Operation of a pnp transistor

The symbol for a pnp transistor (Figure 2.16) is identical to that of an npn type except that the arrow on the emitter shows conventional current flowing towards the base. Operation is explained in the same way but from the simple sketch it can be seen that the main current carriers are holes instead of electrons and the batteries are reversed.

With only battery A in circuit, current flow through the transistor is resisted by junction J_2. Here, negative electrons in the base are attracted by the positive terminal of battery A away from the junction; positive holes are pulled away by attraction to the negative battery terminal. The junction is reverse biased.

When battery B is connected between the base and emitter, current flows through J_1 which is forward biased by B. Holes, as positive charge carriers, move towards the junction under the influence of the negative terminal of B. The few negative electrons in the base are affected by the positive terminal of B and also move to junction J_1. The mutual interaction is such that holes appear in the base (minority carriers) and these act as current carriers for flow from battery A. The size of current flow from A through the transistor is governed by the strength of the input voltage signal from B. Variation of this will cause change in current flow through the device from battery A.

Amplification

In the descriptions of npn and pnp transistors, the same sort of circuit was used in each with small signal power from a side circuit B controlling the larger power in a circuit A. The transistor enables a signal, too weak in itself to be of direct use, to control a larger power source (battery A in the examples). The control by a small available power over a large usable power is called power **gain** or **amplification**. Transistors can be connected in different ways and they can be used for various purposes, including switching.

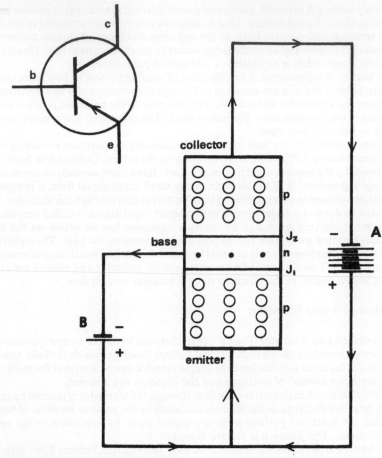

Figure 2.16 *Transistor with pnp arrangement*

Thyristors

There are a number of electronic devices which are classed as thyristors. The most commonly known are **silicon controlled rectifiers** (SCRs) and **triacs**. Like diodes and transistors they have alternate p and n layers; but whereas diodes have two layers and transistors three, silicon controlled rectifiers are four-layer devices and triacs have a greater number.

Thyristors are solid state switches which are turned on by application of a low-level signal voltage through a trigger connection known as a gate electrode. A solid state or static switch has no moving parts to wear, or contacts which can be damaged by arcing. Electrical 'turn on' makes it ideal for remote operation and its small size makes it a convenient component of control circuits. Despite their small size, thyristors can be used to control currents greater than 1000 amps and voltages in excess of 1000 volts. They can therefore replace large conventional switches.

Thyristors operate at a much faster rate than mechanical switches and some are used in services where the switching rate is 25,000 times per second.

Silicon controlled rectifiers (SCRs)

Another name sometimes used for the silicon controlled rectifier is **reverse blocking triode thyristor**.

SCRs are supplied for industrial use in a form which allows them to be easily connected. The type shown (Figure 2.17) has a pnpn pellet of semi-conductor material which is mounted on a short stud and protected by an enclosing metal or ceramic case. The stud is the anode connection, the wire at the other end is the cathode. Circuits to be switched are connected through the anode and cathode. The low-level signal (trigger) voltage is applied at the third lead which is the gate electrode. The SCR with its three connections is represented by an arrow as for a semi-conductor junction rectifier with an extra lead for the gate. It can be used as a rectifier with controlled operation or as a switch. Direction of working current is shown by the arrows.

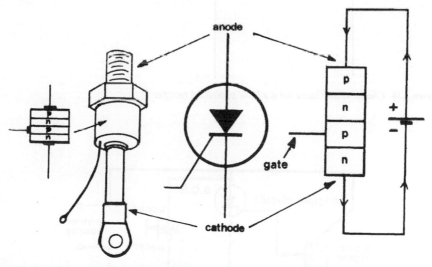

Figure 2.17 *Silicon controlled rectifier (SCR)*

SCR operation

An SCR in circuit will resist current flow in the working direction, apart from leakage (Figure 2.18), because the middle junction is np or reverse biased. Leakage is somewhat greater than with a simple diode because the pn junctions on either side of the centre one are sources of extra minority carriers when forward biased.

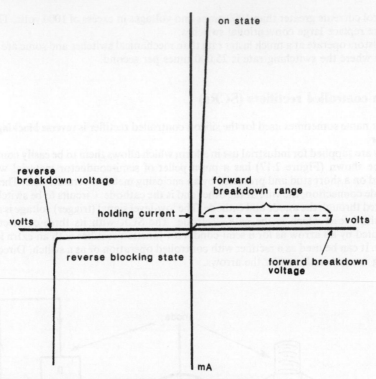

Figure 2.18 *Characteristic curve for a silicon controlled rectifier*

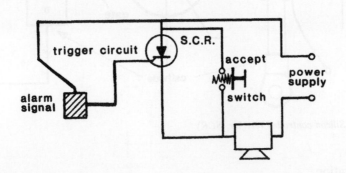

Figure 2.19 *An SCR used in an alarm circuit*

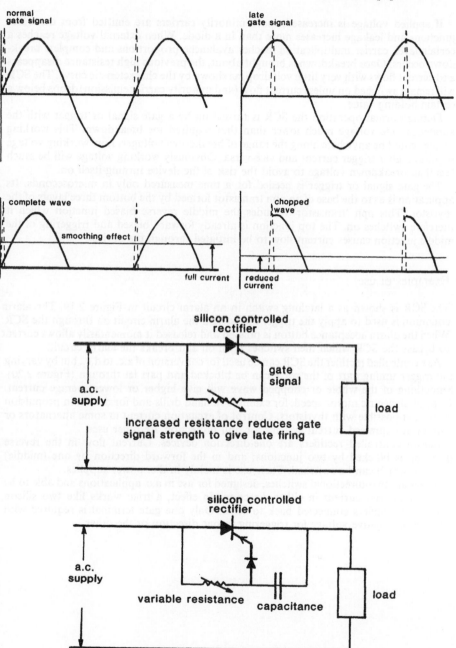

Figure 2.20 *Simple circuits to give late firing of SCR and control over the level of rectified current*

If applied voltage is increased, more minority carriers are emitted from the outer junctions and leakage increases more than in a diode. When external voltage reaches a certain level, carrier multiplication reaches avalanche proportions and complete breakdown occurs. Once breakdown is brought about, the previous high resistance disappears and current flows with very little volt drop, as shown by the characteristic curve. The SCR will remain switched on unless current flow (and minority carrier emission) drops below a certain holding value.

During normal operation the SCR is turned on by a gate signal or trigger with the anode/cathode voltage much lower than that required for breakdown. This working voltage could be anywhere along the range of breakdown voltages. Low working voltage requires higher trigger current and vice-versa. Obviously working voltage will be much less than breakdown voltage to avoid the risk of the device turning itself on.

The gate signal or trigger is needed for a time measured only in microseconds. Its application is as to the base of an npn transistor formed by the bottom three layers of the thyristor. This npn 'transistor' includes the middle reverse biased junction which it therefore switches on. The top junction is already forward biased and triggering of the middle junction causes current flow to be initiated through the device.

Examples of use

The SCR is shown as a latching switch in an alarm circuit in Figure 2.19. The alarm condition is used to apply the trigger and switch the alarm circuit on through the SCR. When the alarm acceptance button is pressed and released it momentarily allows current to bypass the SCR which then switches itself off and breaks the alarm circuit.

As a controlled rectifier the SCR can be used for conversion of a.c. to d.c., but by varying the trigger timing part of the wave can be blocked and part let through (Figure 2.20). Smoothing of the whole or chopped wave will give higher or lower average current. Control of electric motor speed, for example in hand drills and for d.c. main propulsion motors, is possible with thyristors. Control of excitation current in some alternators or current in impressed current systems for hull protection are other uses.

Silicon controlled rectifiers are one-direction devices. Current flow in the reverse direction is blocked by two junctions; and in the forward direction by one (middle) junction which can be switched. Their use is mainly in d.c. power supplies.

Triacs are two-directional switches, designed for use in a.c. applications and able to be switched to pass current in either direction. In effect, a triac works like two silicon controlled rectifiers connected back to back. Only one gate terminal is required with positive or negative voltage for triggering in one direction or the other.

CHAPTER THREE

A.C. Generators

The operation of generators relies on the principle that whenever there is mutual cutting between a conductor and a magnetic field, a voltage and resulting current will be induced in the conductor. The flow of induced current is not random: it is governed by the directions of cutting and of the field and can be found from Fleming's Right Hand Rule (Figure 3.1).

Magnitude of the induced voltage depends on the strength of the magnetic field, rate of cutting and length of the conductor.

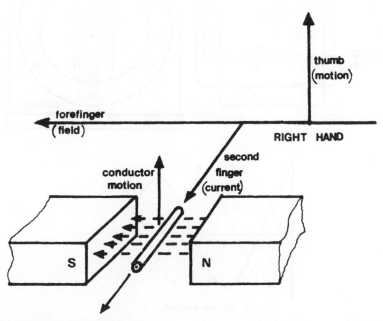

Figure 3.1 *Fleming's Right Hand Rule*

Simple alternator

The arrangement used in the majority of alternators to exploit the principle of generation is shown simply in the sketch (Figure 3.2). Mutual cutting between conductors and magnetic fields is produced by rotating poles, the magnetic fields of which move through fixed conductors.

The rotor shown has a pair of poles so that output is generated simultaneously in two conductors. Reference to the Fleming Right Hand Rule will confirm the instantaneous direction of conventional current indicated by the arrows. The two conductors (R and R_1) are connected in series so that the voltages generated in them add together to deliver current to the switchboard. The rotating fields, although moving at constant speed, will cut the conductors at a changing rate because of the circular movement. Voltage induced at any instant is proportional to the sine of the angle of the rotating vector. The pattern of build-up and decline, and also reversal in the voltage induced, is shown by the sine wave R.

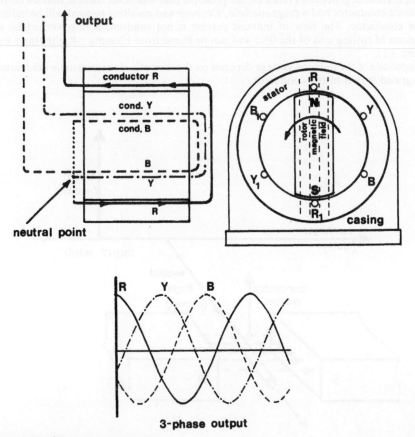

Figure 3.2 *Simple practical alternator*

Voltage and current are generated in each of the pairs of conductors in turn – first in one direction and then in the other – to produce three-phase alternating current. The effect in conductors Y and B is also shown.

Three-phase systems

Outputs from the three sets of conductors in the alternator are delivered to three separate bus-bars in the switchboard. This is necessary because of the voltage and current disparity between them at any instant.

Three-phase, four-wire systems use a single return wire which is connected to the neutral point of the star windings. Current in the return wire is the sum of currents in the individual phases. If loads on each phase are balanced with voltages equal and at 120° apart, the three currents will sum to zero and the return wire will carry no current. The fourth (return) wire will carry a small current if there is imbalance.

Three-phase, three-wire systems have no return wire. This is acceptable for ships where, direct from the main switchboard, three-phase motors make up much of the load and unless there is a fault they take current equally from the phases. Also some imbalance is acceptable with a three-phase, three-wire system provided load is connected in delta. Supplies for lighting, heating, single-phase motors and other loads are taken through delta–star or delta–delta transformers.

The neutral point

The majority of British ships use three-phase, three-wire distribution with the neutral points of alternators insulated (Figure 3.3). Very little current will flow through an earth fault on one phase, because there is no easy path for it back to the electrical system. With such a system, an essential electric motor with an earth fault can be kept running until stoppage for repair is convenient. This would be as soon as possible to avoid a full-phase fault that would result if an earth occurred on another phase as well.

Although fault current is negligible with an insulated/unearthed neutral point, overvoltages are high. The transient likely is 2.5 × line voltage. Such a voltage surge is within the capability of the main insulation of marine electrical equipment which is tested to 2 × line voltage + 1000 volts.

A few British vessels have electrical distribution systems with an earthed neutral (Figure 3.4). This is a connection of the system, via the neutral point of the alternator, to the hull steel. The result of not isolating the electrical system from the hull is that current flow from an earth fault on any phase has a path through the hull steel and earthed neutral back to the system. The availability of the path encourages higher fault current flow than is the case where the neutral is insulated or connected by resistance. Equipment with an earth fault, where the system is earthed, must be disconnected immediately if a fault develops. This can be effected automatically with an earthed neutral system because the level of fault current is high enough to operate a trip.

Earth fault current is high with earthed neutral systems, but overvoltages due specifically to earth faults are lower. The earthed system is chosen to limit overvoltages and to give automatic earth fault location and disconnection.

Overvoltages due to switching are not affected by choice of earthing or insulating the neutral. These high surges, and the possibility of others from failure of the voltage regulator, mean that the same standard of equipment insulation is required for both arrangements.

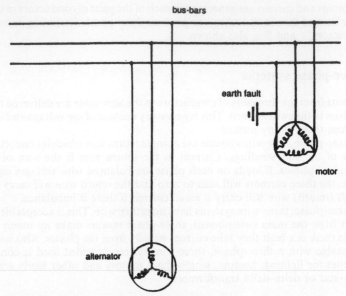

Figure 3.3 *Three-phase three-wire system with insulated neutral*

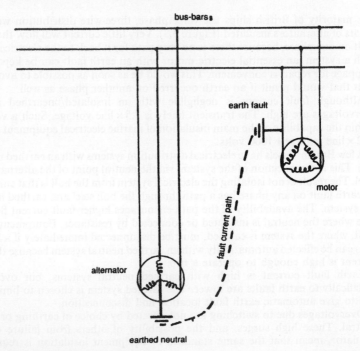

Figure 3.4 *Distribution with earthed neutral*

Stator construction

The stator is a tube made up of silicon iron laminations with axial slots along the inside surface (Figure 3.5) in which the conductors are laid. The importance of the iron stator lies in its ability to strengthen the rotor magnetic fields which cut the conductors. Obviously the iron stator is also a conductor and will have, like the conductors, voltage and current induced in it by the rotating fields. It is to prevent circulation of unwelcome eddy currents that the stator is made up as a laminated structure of steel stampings.

For assembly, the slotted laminations (each insulated on one side) are built into a pack with a number of distance pieces. Substantial steel endplates welded to external axial bars serve to hold the laminations firmly. The distance pieces are inserted to provide radial ventilation ducts for cooling air.

High conductivity copper in the form of round wire, rectangular wire or flat bar is used for the conductors. Round or rectangular wire used in smaller machines is coiled into semi-enclosed and insulated slots (Figure 3.6). This type of slot improves the magnetic field.

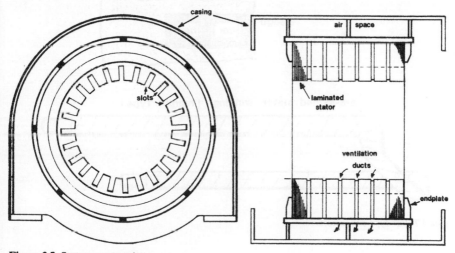

Figure 3.5 *Stator construction*

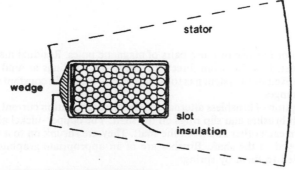

Figure 3.6 *Cross-section of stator slot and winding*

Rectangular section copper bar conductors in large alternators are laid in open slots (Figure 3.7). The bars are insulated from each other and from the metal slot surfaces by a mica-based paper and tape cladding. Wedges are fitted to close the slots and retain the windings. In some machines the wedges are of magnetic material which helps to make the field more uniform, so reducing pulsation and losses. Bonded fabric wedges are used in some alternators. Slots may be skewed to reduce pulsation and waveform ripple.

Insulating materials used in the slots and around the conductors are porous. A method of sealing is necessary to exclude moisture which would cause insulation breakdown. The sealing procedure starts with drying the stator with its assembled conductors. The windings and insulation are then impregnated with synthetic resin or varnish and oven-cured.

The six-conductor winding in Figure 3.2 shows the principle of three-phase generation. The far greater number of windings in an actual machine require much more complication in the winding.

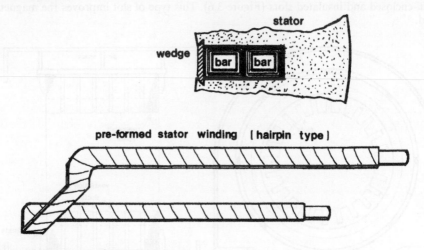

Figure 3.7 Bar-type conductors

Rotor details

An alternator rotor has one or more pairs of magnetic poles. Residual magnetism in the iron cores is boosted by flux from direct current in the windings around them and this current from the excitation system has to be adjusted to maintain constant output voltage through load changes.

With the exception of brushless alternators, the direct excitation current for the rotor is supplied through brushes and slip rings on the shaft. The copper–nickel alloy rings must be insulated from each other and from the shaft. They are shrunk on to a mica-insulated hub which is keyed to the shaft. Brushes are of an appropriate graphite material and pressure is applied to them by springs.

Cylindrical rotor construction

The cylindrical rotor (Figure 3.8) is constructed with axial slots to carry the winding, which forms a solenoid although not of the usual shape. Direct current from the excitation system produces a magnetic field in the winding and rotor so that N/S poles are formed on the areas without slots. One rotation of the single pair of poles will induce one cycle of output in the stator windings (conductors). An alternator with one pair of poles has to rotate at 60 times per second to develop a frequency of 60 cycles per second. In terms of revolutions per minute, the alternator speed must be $60 \times 60 = 3600$ r.p.m.

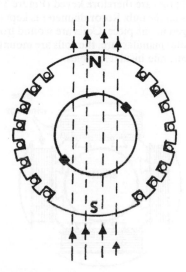

Figure 3.8 *Cylindrical or turbo-alternator rotor*

Projecting (salient) poles bolted to the periphery of a high-speed rotor would be subject to severe stress as the result of centrifugal force. The effect is minimised by using the cylindrical type of construction with the poles being built into the rotor. Small diameter is compensated for by length.

Alternators with one pair of rotor poles are designed for steam or gas turbine drive through reduction gears. For this reason they are sometimes referred to as turbo-alternators. Cylindrical rotors can also be wound with two pairs of poles.

Construction of the core is similar in principle to that of the stator. It is built up of steel laminations pressed together between clamping rings and keyed to the steel shaft. Ventilating ducts are formed by spacers. The semi-enclosed slots are lined with mica-based insulation and the insulated copper windings are retained by phosphor–bronze wedges. The end turns of the winding are held against centrifugal force by rings of insulated steel wire or other material. After winding, rotors are immersed in resin or varnish and oven-cured.

42   A.c. generators

Salient pole rotor construction

Salient field poles are those which are secured to the periphery of the alternator rotor and therefore project outwards. The word 'salient' describes only the physical construction of the rotor: it is not an electrical term.

Higher speeds

A rotational speed of 1800 r.p.m. in an alternator with two pairs of poles which is designed for a supply frequency of 60 Hz (cycles per second) produces severe stress as the result of centrifugal force. The solid poles are therefore keyed (Figure 3.9) or firmly bolted (Figure 3.15) to flat machined faces on the hub. Rotor diameter is kept to a minimum and the mild steel hub and shaft are forged in one piece. Coils are wound from copper strip interleaved with insulating material. After manufacture, the coils are mounted between flanges of hard insulating board on the micanite insulated poles.

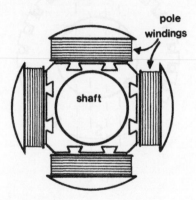

Figure 3.9 *High-speed salient pole rotor*

Lower speeds

Alternators designed for rotation at lower speeds with slower prime movers have a greater number of pairs of poles (Figure 3.10). One pair of poles induces only one complete cycle in the stator windings per revolution, and where for example an engine-driven alternator is intended for operation at 600 r.p.m. it requires six pairs of poles. The number of pairs of poles, p, is found from proposed speed and system frequency

$$p = \frac{f\,(\text{frequency}) \times 60}{N\,(\text{r.p.m.})}$$

Frequency, f, is in cycles per second or hertz (Hz), so the figure is multiplied by 60 to make it cycles per minute in the calculation.

The laminated poles of multi-pole machines are riveted and strengthened for bolting to the rotor hub by a mild steel bar inserted in the steel laminations. The coils are of insulated copper strip wound on mica-insulated spools of galvanised steel. Spools with their

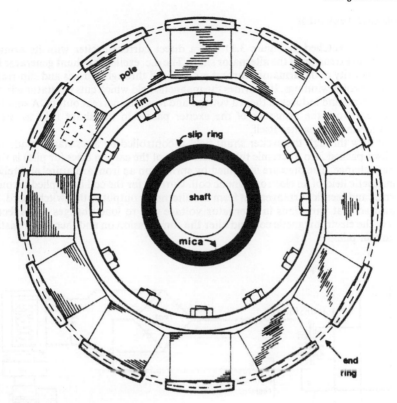

Figure 3.10 *Slow-speed salient pole rotor*

windings are fitted onto the poles and the assembly is bolted to the machined surface on the hub. Laminated poles have a damper winding which consists of copper bars in the pole faces, joined by sectionalised copper end rings.

Rotor diameter is larger on slow-speed machines to accommodate the large number of poles, but low rotational speed produces less stress from centrifugal force. The flywheel effect is beneficial and this, together with careful balance, improves smooth running.

Excitation systems

The excitation system has both to supply and control the direct current for the rotor pole windings. Level of the excitation current and resulting pole field strength are automatically adjusted by the voltage regulating component. Excitation of early alternators was provided by a small direct current generator; voltage control by a carbon pile regulator. This, often referred to as the conventional alternator, is described below together with the vibrating contact, brushless and self-excited types.

Carbon pile regulator

The alternator sketched in Figure 3.11 has a direct current exciter with its armature mounted on an extension of the alternator shaft. The d.c. exciter is a shunt generator from which the majority of the armature current is conveyed through brushes and slip-rings to the alternator rotor windings. It provides the magnetic field which cuts the stator windings as the rotor turns and induces in them a voltage and resulting current output. A small part of the current from the armature of the exciter passes to the shunt field to provide excitation for the d.c. exciter itself.

Current flow through the exciter shunt field is controlled by a resistance made up of carbon discs packed into a ceramic tube. Resistance of the carbon discs (or pile) is varied by pressure change. The pressure is applied by springs on an iron 'armature' and relieved by the magnetic field of an electromagnetic coil. Current for the coil is supplied through a transformer and rectifier arrangement from the alternator output to the switchboard. This is designed so that variations in alternator voltage due to load changes will affect the strength of the electromagnetic coil and alter the compression on and therefore resistance of the carbon pile.

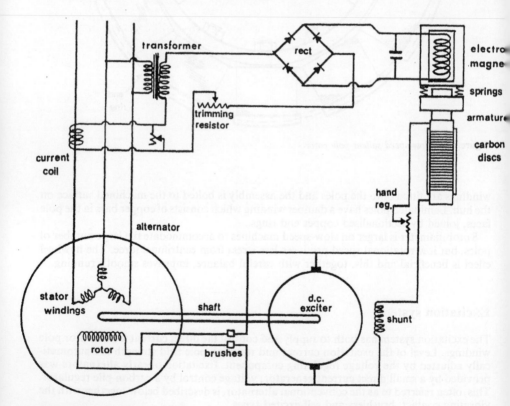

Figure 3.11 *Conventional alternator with DC exciter and carbon pile regulator*

The resistance of carbon is least when pressure exerted on it by the springs and armature is greatest, and this occurs when low alternator output voltage causes the solenoid to be weak. With low resistance between armature and shunt field of the d.c. exciter, more current flows to the shunt and the high excitation current produced is fed to the alternator rotor and increases the voltage.

Resistance of the carbon pile is highest when pressure on it is reduced by a strong solenoid field when alternator output voltage is high. A strong field pulls the iron armature away from the pile, against the pressure of the springs.

Springs shown in the sketch are of the coil type but those in a carbon pile regulator are leaf springs. The ceramic tubes and discs are fitted in a casing with cooling fins. The solenoid has an iron core which, like the ballast resistance, is set to give the correct characteristics. The trimming resistor and hand regulator are for adjustment of the initial setting of the regulator.

Vibrating contact regulator

The operating coil of a vibrating contact regulator is similar to that of the carbon pile type. It is supplied with a transformed and rectified current from the alternator output (Figure 3.12), and the field of the electromagnet is used to attract an iron armature against a spring. The spring acts as a voltage reference and the strength of the electromagnet is related to alternator output voltage. Increase of output voltage produces an increase in strength of the magnet, and the effect is that the plunger and lever are pulled up together

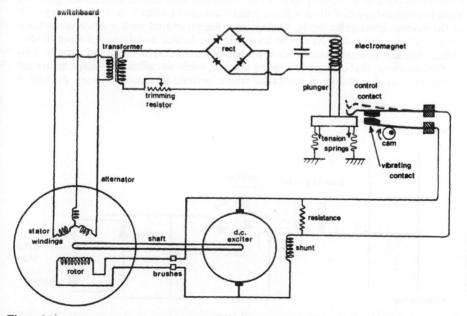

Figure 3.12 *Alternator and DC exciter with vibrating contact AVR (greatly simplified)*

with the control contact. A drop in voltage and decrease of strength of the coil allows the plunger to be pulled downwards by the spring so that the control contact moves down also. The position of the control contact is governed by alternator voltage through the coil.

While the control contact will move with voltage variation, the vibrating contact is vibrated continuously by a rotating cam at a rate of 120 times per second. The vibrating contact touches the other during the upper part of each vibration and the control contact is free to be moved up by the other, being only in spring contact with the plunger level. High alternator output voltage, through the coil, causes the control contact to be lifted so that the vibrating contact only touches briefly. Low output voltage and weaker field in the coil allows the plunger to drop so that the contact time is longer.

Closing of contacts has the effect of short-circuiting the resistance in series with the shunt field of the d.c. exciter. Shunt current passes through the contacts in preference to passing through the resistor and the larger shunt current increases flux and output of the exciter. This in turn provides extra excitation and increases alternator output voltage.

Static automatic voltage regulator

The carbon pile regulator uses a magnetic coil powered from the alternator output. Strength of the field varies with alternator voltage and this strength is tested against springs which are the voltage reference. The moving contact regulator employs a similar matching of alternator output effect through a magnetic coil against springs.

The availability of a small transformed, rectified and smoothed power supply from the alternator output makes possible the matching of it directly against an electronic reference in the static automatic voltage regulator. The direct current derived from the alternator output is applied to a bridge (Figure 3.13) which has fixed resistances on two arms and variable resistances (zener diode voltage references) on the other two. The zeners operate in the reverse breakdown mode, having been manufactured with a zener breakdown voltage of very low value. As can be seen from the earlier description of zener diodes, voltage remains constant once breakdown has occurred despite change in current. This implies, however, that changes in applied voltage, while not affecting voltage across the diode, will cause a change in resistance which permits change in current. As with a

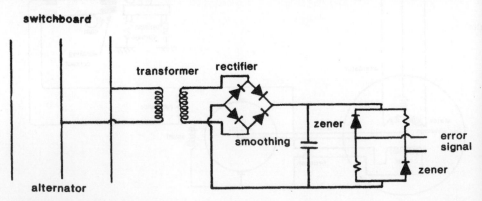

Figure 3.13 *Static automatic voltage regulator (SAVR)*

Wheatstone bridge, imbalance of the resistances changes the flow pattern and produces in the voltage measuring bridge an error signal.

The error signal can be amplified and used to control alternator excitation in a number of different ways. Thus it can control the firing angle of thyristors (Figure 3.14) through a triggering circuit to give the desired voltage in the brushless alternator described. It can be used in the statically excited alternator to correct small errors through a magnetic amplifier arrangement. The error signal has also been amplified through transistors in series, for excitation control.

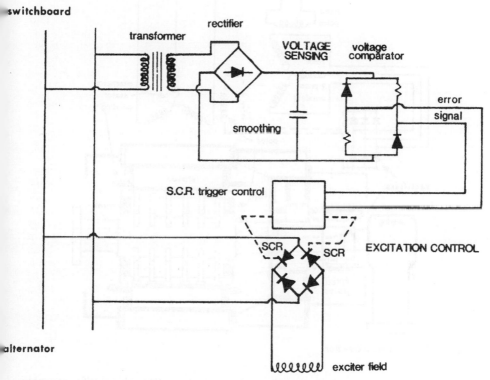

Figure 3.14 *Error signal used to control thyristors in the excitation system*

The brushless alternator

In this machine slip-rings and brushes are eliminated and excitation is provided not by conventional direct current exciter but by a small alternator. The a.c. exciter (Figure 3.15) has the unusual arrangement of three-phase output windings on the rotor and magnetic poles fixed in the casing. The casing pole coils are supplied with direct current from an automatic voltage regulator of the type described in the previous section. Three-phase

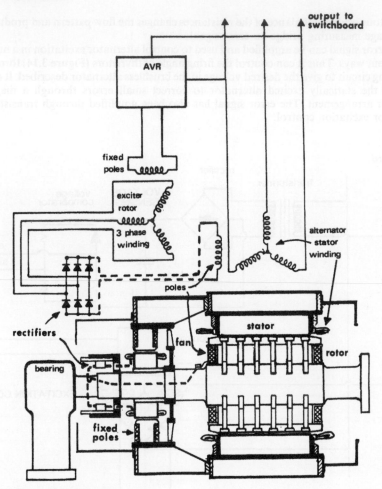

Figure 3.15 *Brushless alternator*

current generated in the windings on the exciter rotor passes through a rectifier assembly on the shaft and then to the main alternator poles. No slip-rings are needed.

The silicon rectifiers fitted in a housing at the end of the shaft are accessible for replacement and their rotation assists cooling. The six rectifiers give full-wave rectification of the three-phase supply.

Static excitation system

Direct on-line started induction motors take six to eight times the normal full load current as they are starting. A large motor therefore puts a heavy current demand on the a.c. system, causing voltage to dip and, where recovery from the dip is slow, also producing momentary dimming of lights and effects on other equipment. There is a limit to the size of motor that can be started direct on-line, but the ability of alternators to recover from large starting currents has been enhanced by development of the static excitation system.

The direct current required for production of the rotor pole magnetic field is derived from alternator output without the necessity for a rotating exciter as described for the carbon pile/d.c. exciter system or for the brushless machine.

The principle of the static or self-excitation system (Figure 3.16) is that a three-phase transformer with two primaries, one in shunt and the other in series with alternator output, feeds current from its secondary windings through a three-phase rectifier for excitation of the main alternator rotor.

Excitation for the no-load condition is provided by the shunt-connected primary, which is designed to give sufficient main rotor field current for normal alternator voltage at no

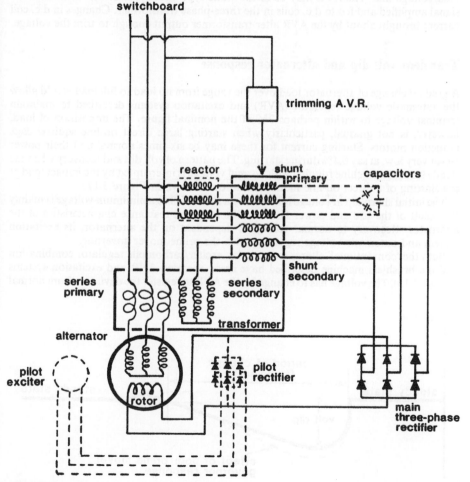

Figure 3.16 *Static excitation system principle*

load. Reactor coils give an inductive effect so that current in the shunt winding lags main output voltage by 90°. Build-up of voltage at starting is assisted by capacitors which promote a resonance condition with the reactors or by means of a pilot exciter (these alternatives are shown on the sketch by broken lines).

Load current in the series primary coils contributes the additional input to the excitation system to maintain voltage as load increases. Variations in load current directly alter excitation and rotor field strength to keep voltage approximately right.

Both shunt and series inputs are added vectorially in the transformer. Diodes in the three-phase rectifier change the alternating current to direct current which is smoothed and fed to the alternator rotor through slip-rings.

Voltage control within close limits is achieved by trimming with a static AVR to counteract small deviations due to internal effects and wandering from the ideal load/voltage line. The AVR may be of the static type already described with the error signal amplified and fed to d.c. coils in the three-phase transformer. Changes in d.c. coil current brought about by the AVR alter transformer output enough to trim the voltage.

Transient volt dip and alternator response

A gradual change of alternator load over the range from no load to full load would allow the automatic voltage regulator (AVR) and excitation systems described to maintain terminal voltage to within perhaps 2% of the nominal figure. The imposition of load, however, is not gradual, particularly when starting large direct on-line squirrel cage induction motors. Starting current for these may be six times normal and their power factor very low, at say 0.4% during starting. The pattern of volt dip and recovery when the steady state of a machine running with normal voltage is interrupted by the impact load at the starting of a direct on-line induction motor is shown in Figure 3.17.

The initial sharp dip in voltage followed by a slower fall to a minimum voltage is mainly the result of the size and power factor of the load and reactance characteristics of the alternator. Recovery to normal voltage is dependent on the alternator, its excitation system and automatic voltage regulator; also the prime mover governor.

Both the 'conventional' alternator with d.c. exciter/carbon pile regulator combination and the brushless machine described have error-operated AVR and excitation systems (Figure 3.18). The voltage has to change for the AVR to register the deviation from normal

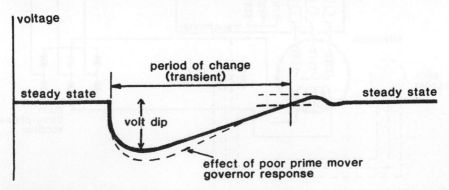

Figure 3.17 *Typical volt dip/recovery pattern for an alternator*

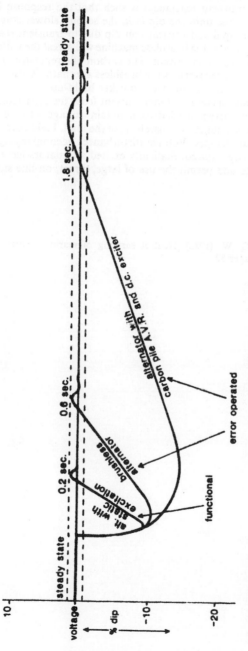

Figure 3.18 *Comparisons of volt dip and recovery for different excitation systems*

and to then adjust the excitation for correction. The suddenness of the initial volt dip (blamed specifically on transient reactance) is such that the response from the error-operated system cannot come until the dip is in the second slower stage. Thus neither machine can prevent the rapid and vertical volt dip due to transient reactance, but the faster acting voltage regulator of the brushless machine will arrest the voltage drop sooner on the slower secondary part of its descent. The carbon pile regulator is slow compared with the static type but better recovery by the brushless alternator is also achieved by field forcing, i.e. boosting the excitation to give a quicker build-up.

Static excitation systems make use of load current from the alternator to supply that component of excitation current needed to maintain voltage as load increases. This component of excitation is a 'function', therefore, of the load. Field current is thus forced to adjust rapidly as load changes. Voltage disturbances accompanying application or removal of load are greatly reduced. Statically excited alternators have better recovery from voltage disturbance and permit the use of large, direct, on-line starting induction motors.

Reference

Ball, R., and Stephens, G. W. (1982) 'Neutral earthing of marine electrical power systems'. *Trans.I.Mar.E.* **95**, Paper 32.

CHAPTER FOUR

A.C. Switchboards and Distribution Systems

Deadfront switchboards are a safety requirement for a.c. voltages in excess of 55 V. Mechanical strength and a non-flammable construction are obtained with the use of sheet steel for the cubicles, and a passageway of at least 0.6 m is left at the rear for access.

Power factor

The power of a direct current system is found by multiplying together the readings of voltage and current (V × A = watts). The same procedure with an a.c. system will produce a figure which is contradicted by the lower wattmeter readings of true power. The ratio between true and apparent power is termed the **power factor**.

$$\text{Power factor} = \frac{\text{true power}}{\text{apparent power}} = \frac{\text{watts}}{\text{volt-amps}} = \frac{\text{kW}}{\text{kVA}}$$

The anomaly between d.c. and a.c. systems is the result of the continual change in strength and direction of alternating current. Build-up of current is accompanied by a build-up of magnetic field which, as it cuts through the windings of alternators and motors in the installation, generates a voltage which reacts against the main voltage. The effect of this inductive reactance is to cause the supply current to lag the main voltage. Lateness of the current means that part of it is useful and part is not. The useful part is registered on the wattmeter as **true power**, sometimes called the **wattful component**. The power factor figure (usually about 0.8) is also the cosine of the angle of lag of the current.

Only the true power output and losses have to be supplied by the prime mover. However, the sizes of alternators, motors and transformers are governed by the apparent power.

Alternator circuit breakers

The air-break circuit breakers used for marine installations are frame-mounted and arranged for isolation from busbar and alternator input cable contacts by being moved horizontally forward to a fixed position. The isolating plugs (Figure 4.1) are not designed for making or breaking contact on load so the breaker must be open before the assembly is withdrawn. Safety interlocks should be fitted to ensure that the circuit breaker assembly

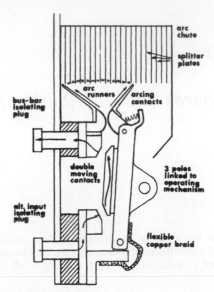

Figure 4.1 *Air-break circuit breaker contacts*

cannot be racked out or in with contacts closed. Maintenance is facilitated by the draw-out compartments and extending guide rails which allow the breaker to be pulled out completely.

The alternator breaker for three-phase supply has a single unit for each phase, of similar design to the example in Figure 4.1. The three units are linked together by an insulated bar for simultaneous operation. Main fixed and moving contacts are constructed of high-conductivity copper, and as an aid to low contact resistance the faces are silver plated. Main contacts are designed to carry normal full load current without overheating and overload current until tripped, when a fault occurs.

Interruption of current flow results in the production of an arc between contact faces. Arcing is severe with overload current but is not a serious problem during normal operation. To prevent damage to main contacts, separate arcing contacts are fitted which are designed to open after and close before main contacts. These supplementary contacts are of arc-resisting alloy such as silver tungsten and easily replaced if damaged. They should be inspected after fault operation of the breaker. The arcing contact shown has a spring which pushes it forward to hold until after main contacts have opened.

Air-break circuit breakers with arcing contacts have always been used for d.c. switchboards but further development was necessary to improve arc control before such breakers could replace oil circuit breakers and be used for a.c. (pre-1940). Later development has meant that they can be used on systems with voltages up to 3.3 kV.

Arc control requires that the arc be elongated and removed from the gap between the arcing contacts. Electromagnetic forces associated with the arc and thermal action cause it to move up the arc runners to the arc chute provided for the purpose. Thus the arc is elongated and finally chopped into sections and cooled by the splitter plates.

Arc chutes are of insulating and arc-resisting material. They confine the arc and produce a funnel effect which assists thermal action. Splitter plates are of metal (steel or copper) in some breakers, and in others of insulator material. Some breakers have horizontal rods fitted to cool and split the arc. Arc runners are fixed and not of the moving type in some designs.

Interruption of the arc is assisted by the current dropping to zero during the cycle (however, with three-phases the zero points in each phase are staggered). Contact opening is therefore followed by a current zero and this means that for the next part of the cycle, an arc has to be struck across a gap. Successful removal of ionised gas (from the arc which resulted from contact opening) will increase resistance in the air gap between contacts. When gas remains, it provides a path across which the arc can re-strike. The rate at which the gas is removed is such that the arc will not re-strike more than two or three times.

Breaking speed is made as high as possible by powerful throw-off springs and light construction of the moving arm assembly. Rebound at the end of the opening movement is prevented by anti-bounce devices.

Rapid closing of the breaker also helps to prevent damage and most are power, rather than manually, closed. Power is provided by a solenoid or by a spring which is automatically rewound by a motor and left ready after each closing operation. Where springs are used, an emergency hand-tensioning method is arranged for use with a dead board, so that the spring can be wound up ready for closing the breaker. After operation of a spring-activated breaker, the rewind motor can usually be heard charging the spring for the next time.

Alternator and system protection

Protective devices are built into main alternator breakers to safeguard both the individual alternator and the distribution system against certain faults. Overcurrent protection is by relays which cut power supplies to non-essential services on a preferential basis, as well as breaker overload current trips and instantaneous short current tripping. A reverse power trip is fitted where alternators are intended for parallel operation (in some vessels they are not), unless equivalent protection is provided by other means. Parallel operation of alternators also requires an under-voltage release for the breaker.

Overcurrent protection

The alternator breaker has an overcurrent trip, but a major consideration is that the supply of power to the switchboard must be maintained if possible. The breaker therefore is arranged to be tripped instantly only in the event of high overcurrent such as that associated with short circuit. When overcurrent is not so high, a delay with inverse time characteristic allows an interval before the breaker is opened. During this time the overload may be cleared.

Overload of an alternator may be due to increased switchboard load or to a serious fault causing high current flow. Straight overload (apart from the brief overload due to starting of motors) is reduced by the preference trips which are designed to shed non-essential switchboard load. Preference trips are operated by relays set at about 110% of normal full load. They open the breakers feeding ventilation fans, air conditioning equipment etc. The non-essential items are disconnected at timed intervals, so reducing alternator load. A serious fault on the distribution side of the switchboard should cause the appropriate supply breaker to open, or fuse to operate, due to overcurrent. Disconnection of faulty equipment will reduce alternator overload.

Inverse definite minimum time (IDMT) relay

Accurate inverse time delay characteristics are provided by an induction type relay with construction similar to that of a domestic wattmeter or reverse power relay.

Current in the main winding (Figure 4.2) is obtained through a current transformer from the alternator input to the switchboard. (The main winding is tapped and the taps brought out to a plug bridge for selection of different settings.) Alternating current in the main winding on the centre leg of the upper laminated iron core produces a magnetic field that in turn induces current in the closed winding. The magnetic field associated with the closed winding is displaced from the magnetic field of the main winding and the effect on the aluminium disc is to produce changing eddy currents in it. A tendency for the disc to rotate is prevented by a helical restraining spring when normal current is flowing. Excessive current causes rotation against the spring and a moving contact on the spindle comes in to bridge, after a half-turn, the two fixed contacts, so that the tripping circuit is closed.

Speed of rotation of the disc through the half-turn depends on the degree of overcurrent. Resulting inverse time characteristics are such as shown in Figure 4.3. In many instances of overcurrent, the IDMT will not reach the tripping position as the excess current will be cleared by other means. The characteristic obtained by the relay is one with a definite minimum time and this will not decrease regardless of the amount of overcurrent. Minimum time, however, can be adjusted by changing the starting position of the disc.

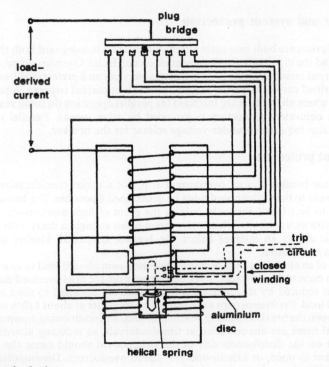

Figure 4.2 *Overload relay*

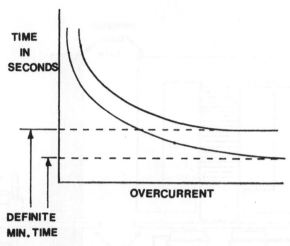

Figure 4.3 *Relay inverse time characteristics*

Alternator breakers have instantaneous short-current trips in addition to IDMT (or other type) relays. In the event of very large overcurrent these rapidly trip the breaker out. Without an instant trip, high fault current would continue to flow for the duration of the minimum time mentioned above.

Reverse power protection

Alternators intended for parallel operation are required to have a protective device which will release the breaker and prevent motoring if a reversal of power occurs. Such a device would prevent damage to a prime mover which had shut down automatically due to a fault such as loss of oil pressure. Reversal of current flow cannot be detected with an alternating supply but power reversal can, and protection is provided by a reverse power relay, unless an acceptable alternative protective device is fitted.

The reverse power relay is similar in construction to a household electricity supply meter (Figure 4.4). The lightweight non-magnetic aluminium disc, mounted on a spindle which has low-friction bearings, is positioned in a gap between two electromagnets. The upper electromagnet has a voltage coil connected through a transformer between one phase and an artificial neutral of the alternator output. The lower electromagnet has a current coil also supplied from the same phase through a transformer.

The voltage coil is designed to have high inductance so that current in the coil lags voltage by an angle approaching 90°. Magnetic field produced by the current similarly lags the voltage and also lags the magnetic field of the lower electromagnet. Both fields pass through the aluminium disc and cause eddy currents.

The effect of the eddy currents is that a torque is produced in the disc. With normal power flow, trip contacts on the disc spindle are open and the disc bears against a stop. When power reverses, the disc rotates in the other direction, away from the stop, and the contacts are closed so that the breaker trip circuit is energised. A time delay of 5 seconds prevents reverse power tripping due to surges at synchronising. Reverse power settings are 2 to 6% for turbine prime movers and 8 to 15% for diesel engines.

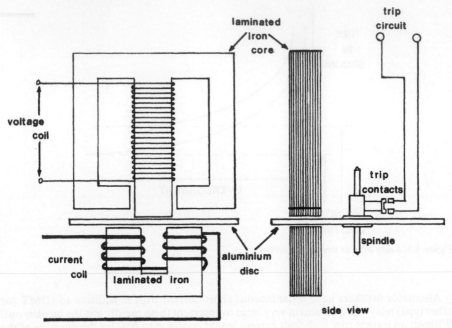

Figure 4.4 *Reverse power relay*

Under-voltage protection

Closure by mistake of an alternator breaker when the machine is dead is prevented by an under-voltage trip. This protective measure is fitted when alternators are arranged for parallel operation. Instantaneous operation of the trip is necessary to prevent closure of the breaker. However, an under-voltage trip also gives protection against loss of voltage while the machine is connected to the switchboard. Tripping in this case must be delayed for discrimination purposes, so that if the volt drop is caused by a fault then time is allowed for the appropriate fuse or breaker to operate and voltage to be recovered without loss of the power supply.

Synchroscope

The operation of synchronising an alternator before paralleling with another machine could be carried out with the type of synchroscope shown in Figure 4.5. With its use, two phases of the incoming machine can be matched with the same two switchboard phases.

The synchroscope is a small motor with coils on the two poles connected across red and yellow phases of the incoming machine and the armature windings supplied from red and yellow switchboard bus-bars. The latter circuit incorporates a resistance and an inductance coil in parallel. The inductance has the effect of delaying current flow through itself by 90° relative to current in the resistance. The dual currents are fed via slip-rings to the two armature windings and produce in them a rotating magnetic field.

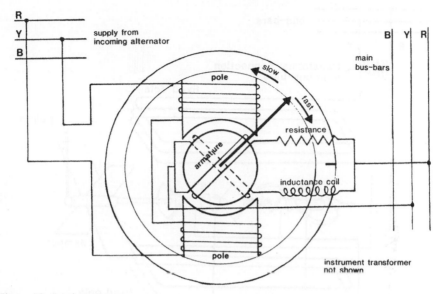

Figure 4.5 *Synchroscope*

Polarity of the poles will alternate north/south with changes in red and yellow phases of the incoming machine, and the rotating field will react with the poles by turning the rotor clockwise or anticlockwise. Direction is dictated by whether the incoming machine is running too fast or too slow. Normal procedure is to adjust alternator speed until it is running very slightly fast and the synchroscope pointer turning slowly clockwise. The breaker is closed just before the pointer reaches the twelve o'clock position, at which the incoming machine is in phase with the switchboard bus-bars.

Another type of synchroscope (Figure 4.6) also uses the principle of resistance and inductance connected in parallel across two alternator phases to give a 90° lag in current flow. The result is that a magnetic field is produced in the coils A and B in turn, first in one direction and then in the other. The pairs of iron sectors are magnetised by the coils through the spindles which act as cores.

The spindle and iron sectors magnetised by coil A, which is supplied through the resistance, have a magnetic field in step with voltage and current of the incoming alternator. This is because pure resistance does not give current a lag, as does the inductance in the circuit for coil B which makes current (and magnetic field) lag voltage by 90°. The iron sector pairs, spindles and coils A and B are separated by a non-magnetic distance piece.

The large fixed poles above and below the spindle are connected across two switchboard bus-bars (the same phases as those in the alternator supplying the spindle coils). When the field of coil A (and the incoming machine) is in phase with the bus-bars, the sectors magnetised by A will be attracted – one to the top coil and the other to the bottom – so that the pointer is vertical. This occurs regularly with the pointer rotating clockwise when the incoming machine is running too fast; also when the machine is too slow and the pointer revolving anticlockwise. Adjustment of incoming alternator speed to match the switch-

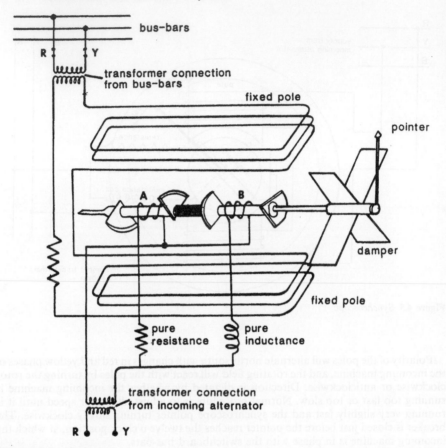

Figure 4.6 *Synchroscope*

board supply frequency results in slower movement of the pointer. Ideally the speed adjustment would achieve a coincidence of phase and speed with the pointer steady at twelve o'clock. In practice, the breaker is closed when the incoming machine is running slightly fast (pointer turning slowly clockwise) and the pointer passing 'five to twelve o'clock'.

Emergency synchronising lamps

The possibility of failure of the synchroscope requires that there is a standby arrangement. A system of lights connected to the switchboard bus-bars and three-phase output of the incoming alternator, shown diagrammatically in Figure 4.7, may be used.

If each pair of lamps were across the same phase the lights would go on and off together when the incoming machine was out of phase with the switchboard and running machine. The alternators would be synchronised when all of the lights were out. Such an

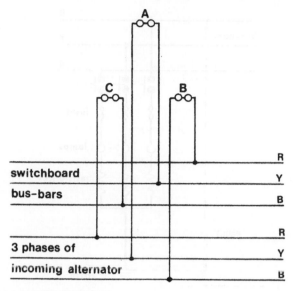

Figure 4.7 *Emergency synchronising lamps*

arrangement is not as good as the one shown where only lamps A are connected across the same phase. Pairs of lamps B and C are cross-connected. At the point when the incoming machine is synchronised, lamp A will be unlit and lamps B and C will show equal brightness. The lamps will give the appearance of clockwise rotation when the incoming machine is running too fast and anticlockwise rotation when it is running too slow.

Pairs of lamps are wired in series because voltage difference between incoming alternator and switchboard varies between zero and twice normal voltage.

A.C. earth fault lamps

The sketch (Figure 4.8) shows the arrangement for earth fault indicator lamps on a three-phase a.c. system. Each lamp is connected between one phase and the common neutral point. Closing of the test switch connects the neutral point to earth. An earth on one phase will cause the lamp for that phase to show a dull light or go out, depending on the severity of the fault.

Each earth lamp and the resistance in series with it provides a path for current flow to the neutral. An earth on one phase will, when the test switch is closed, allow current flow through an easier path than that through the lamp and resistance. The lamp is, therefore, shorted-out and will show a dull light or none at all.

Transformers

The two coils of a simple single-phase transformer are of insulated wire and wound on the same laminated iron core. Often the windings are shown separated as in Figure 4.9 (top).

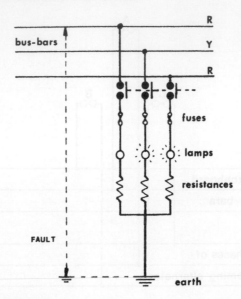

Figure 4.8 A.C. earth fault lamps

Alternating current applied to one winding produces an alternating magnetic flux (strengthened by the iron core) which in cutting the second winding induces a voltage in it; also alternating. The flux, although established in the core by the a.c. source energising the primary winding, also cuts this winding and induces a voltage in it almost equal to the applied primary voltage. The same voltage is induced in each turn of both windings so that voltage is stepped up or down in proportion to the number of turns. The primary winding is always the one connected to the power source; the secondary to the load.

Winding terminals in three-phase transformers are marked with capital letters: $A_1 A_2$, $B_1 B_2$, $C_1 C_2$ for the phases and N for a neutral on the high-voltage (h.v.) winding. Low voltage (l.v.) windings are distinguished by small letters $a_1 a_2$, $b_1 b_2$, $c_1 c_2$ for the phases and n for a neutral. These letters are used regardless of whether the winding is a primary or secondary.

Transformers are incorporated in battery chargers, instrument connections in a.c. work, and power distribution systems. With their use voltage can be stepped up or down by a simple and efficient means without change of frequency. There is no direct link between low-voltage secondary systems and a high-voltage primary. The isolation of sections allows one to be earthed at the neutral and another to be insulated. Short-circuit currents are limited in the secondary. Safety of working supplies can be improved with transformer step-down and isolation from the switchboard.

Three-phase transformers

Primary and secondary windings of three-phase transformers can be connected in star or delta, and various combinations can be used to suit a particular application. Star winding has the advantage of the neutral point which is available for an earth connection if required

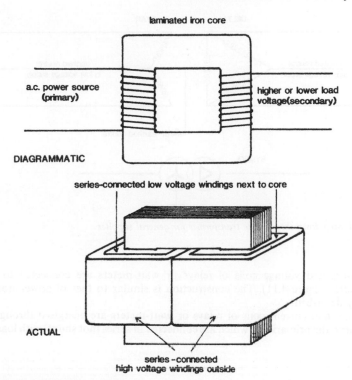

laminated iron core

a.c. power source
(primary)

higher or lower load
voltage(secondary)

DIAGRAMMATIC

series-connected low voltage windings next to core

ACTUAL

series-connected
high voltage windings outside

Figure 4.9 *Simple single-phase transformer*

or for a fourth wire. The delta winding is useful for unbalanced loading but has no neutral point.

A step-down transformer (Figure 4.10) of 440/230 V for general supplies, wound delta–star, would permit earthing of the neutral point for the low-voltage supply with the higher voltage system supplying essential machinery having an unearthed neutral. Earthing of the low-voltage neutral reduces over-voltages when a fault occurs and at the same time causes sufficient fault current to operate protective devices. Essential machinery on the high voltage system with an insulated neutral can be allowed to run if necessary with an earth fault.

Essential machinery sometimes takes its power from high-voltage alternators through a step-down transformer. With the high voltage circuit earthed but the essential machinery requiring to be unearthed, a star–delta transformer can be used.

Instrument transformers

The use of instrument transformers for indirect connection of relays, synchroscopes and measuring instruments means that they are isolated from high voltage and current in main circuits and working with much lower values from transformer secondaries. Thus the instruments are safer and require less insulation.

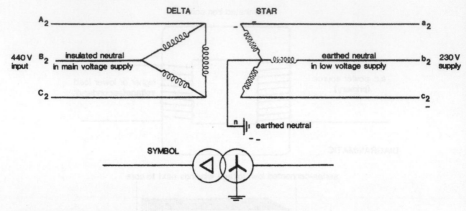

Figure 4.10 *Step-down three-phase transformer for general supplies*

Voltmeters and voltage coils of relays or watt-meters are connected to potential transformers (Figure 4.11). The construction is similar to that of power transformers previously described.

Ammeters and current coils of relays or watt-meters are energised through current transformers, the primaries of which are connected in series (not shunt) with load current.

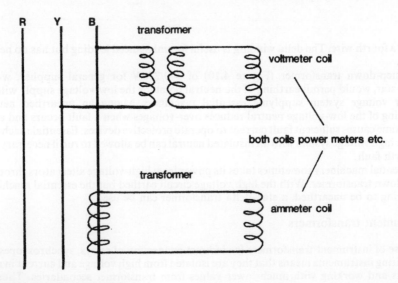

Figure 4.11 *Instrument transformer*

References

Adams, E. M., and Ellwood, E. (1974) 'A 3.3 kV electrical system for large container ships'. *Trans.I.Mar.E.* **86**, Series A Part 9.

de Jong, W., and Robinson, J. N. (1986) 'Generator failures caused by synchronising torques'. *Trans.I.Mar.E.* **99**, Paper 8.

Hennessy, G., McIver, J., Martin, B., and Rutherford, J. (1977) 'Commissioning and operational experience of 3.3 kV electrical systems on Esso Petroleum Products tankers'. *Trans.I.Mar.E.* **89**.

Rush, H. (1982) 'Electrical design concepts and philosophy for an emergency and support vessel'. *Trans.I.Mar.E.* **94**, Paper 28.

Savage, A. N. (1957) 'Developments in marine electrical installations with particular reference to a.c. supply'. *Trans.I.Mar.E.*

Savage, A. N. (1961) 'Details and operating data of recent a.c. installations'. *Trans.I.Mar.E.*

Willcox, A. H. (1969) 'Protective equipment in marine electrical systems'. *Trans.I.Mar.E.*

CHAPTER FIVE

A.C. Motors

The majority of motors on ships with alternating current as the main electrical power are squirrel-cage induction motors with direct on-line starting. With the changeover from the use of direct to alternating current, these motors were a simple and robust replacement for d.c. motors. Compared with d.c. motors (which needed constant maintenance of starting contacts, brushes and commutators; replacement of starting resistances and cleaning) the routine work on an a.c. motor is negligible. Additionally, they are much safer, being non-sparking and having no resistances liable to overheat.

Squirrel-cage induction motors

Three-phase alternating current supplied to the stator windings of the motor when the motor is switched on produces a rotating magnetic field. The field cuts through copper bars in the stationary rotor and induces in them a current. Current flowing through bars and end rings (Figure 5.1) produces a magnetic field in each bar in turn, which in reacting with the rotating field causes the rotor to turn. Motor speed builds up until it almost equals that of the rotating field.

Squirrel-cage rotor construction

Large squirrel-cage rotors have copper bar conductors brazed to copper end rings. In small motors, the bar conductors and end rings may be of aluminium, cast with fan blades on the end rings (Figure 5.2). The bar conductors are arranged in the form of a 'treadmill' type cage (squirrel-cage), but this is not obvious because they are embedded in semi-enclosed slots.

The rotor core is built up of silicon steel stampings individually insulated to eliminate eddy currents and keyed to form a rigid assembly with the shaft. Slots are skewed for smooth starting and quiet running. The purpose of the iron core is to improve magnetic field strength, so the periphery of the rotor is machined accurately to give the smallest possible gap between stator and rotor.

The conductors are a drive fit in the slots to prevent any movement and completed core and cage construction has great strength. The conductor bars may be insulated to cut stray losses.

The mild steel shaft is amply proportioned so that it has the stiffness required to carry

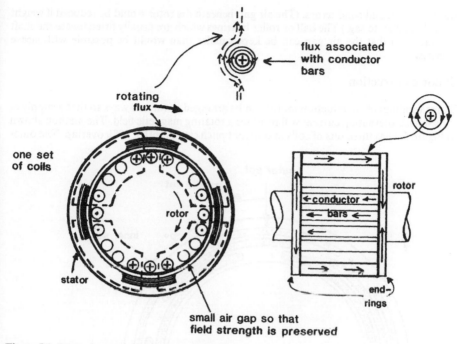

flux associated
with conductor
bars

rotating
flux

one set
of coils

rotor

conductor
bars

end-
rings

stator

rotor

small air gap so that
field strength is preserved

Figure 5.1 *Squirrel-cage induction motor operation*

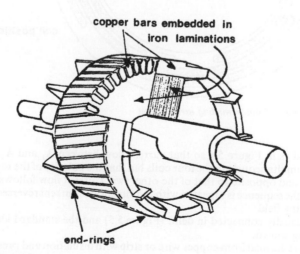

copper bars embedded in
iron laminations

end-rings

Figure 5.2 *Squirrel-cage rotor*

the heavy core and conductors. (The air gap beneath the rotor would be reduced if weight caused the shaft to sag.) The ball or roller bearings which are usually fitted locate the shaft accurately so that the air gap can be kept smaller than would be possible with sleeve bearings.

Stator construction

Stator windings of an induction motor can be arranged in various ways so that a supply of three-phase alternating current will produce a rotating magnetic field. The method shown in Figure 5.3 has three sets of coils at different pitch circles with a 30% overlap. The outer

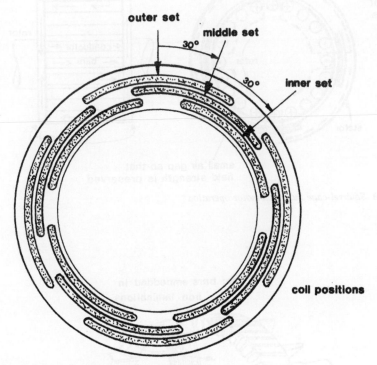

Figure 5.3 *One method of stator winding arrangement*

set are connected as in Figure 5.4 so that current flow between A_1 and A_2 will produce magnetic fields simultaneously in the four coils but that the polarity of the top and bottom will be the same and opposite to that of the other two. Current flow follows in coil sets B and C, and then the sequence is repeated with the direction of current reversed. The effect is to cause a rotating field.

Coil sets are usually connected in delta (Figure 5.5) and the standard identification is with the lettering shown.

The coils are pre-formed from copper wire or strip with insulation and pressed into open slots, or they can be wound into semi-closed slots in the stator core. The stator is made up

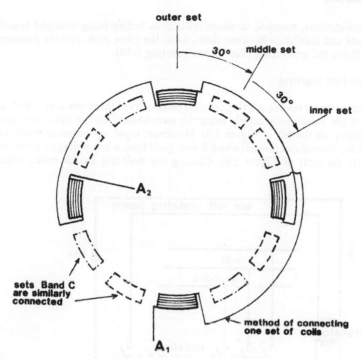

Figure 5.4 *Method of connection of the coils*

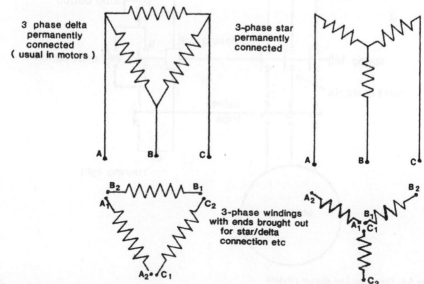

Figure 5.5 *Method of supplying coils*

of steel laminations, stamped to shape with slots before being clamped together. The laminations and insulation between them, as in the rotor core, prevent passage of eddy currents (from the generating effect of the rotating field).

Direct on-line starting

The device for direct on-line starting consists essentially of three contacts, which are closed to connect the three-phase supply from the switchboard to the stator windings of the motor (shown on the left of Figure 5.6). However, rapid operation is beneficial and to achieve this a closing coil is used, which is energised from a low-voltage d.c. control circuit (shown on the right of Figure 5.6). Closing the isolating switch makes main power

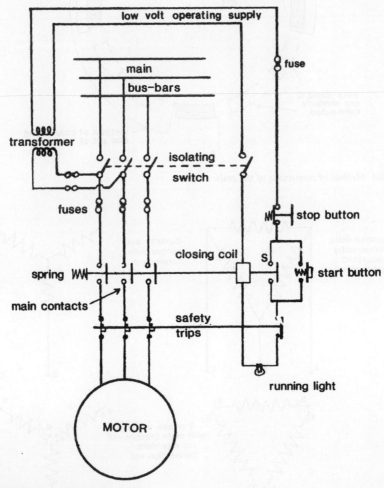

Figure 5.6 *Direct on-line starter circuits*

available to the main contacts and, via the transformer and rectifier, to the operating circuit.

When the start button is pushed the closing coil is energised, and as the main contacts close so also does the contact (S). Release of the start button will not interrupt the closing coil circuit because continuity is maintained through the retaining contact (S).

The main contacts are closed against a spring and de-energising of the closing coil by opening the control circuit with the top button, or operation of safety trips, will cause the contacts to open and the motor to stop. The closing coil acts as a no-volts trip to prevent. involuntary restarting of the motor after a power loss (except for steering gear motors).

Disadvantages

Simple direct on-line start, squirrel-cage induction motors have three disadvantages: (1) high starting current, (2) low starting torque, and (3) single-speed operation (apart from slight slip with increase of torque). Characteristics are shown in Figure 5.7.

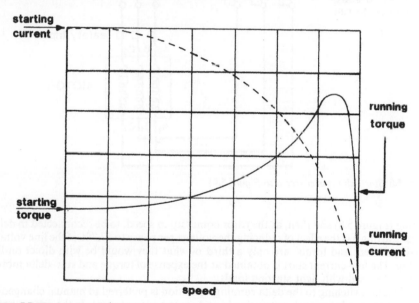

Figure 5.7 *Typical characteristic curves for a direct on-line start squirrel-cage motor*

Low-current starting

Where the high current of direct on-line starting is unacceptable, the squirrel-cage motor can be started by means which reduce voltage and current in the motor stator. A disadvantage of low-current starting is that initial torque is also reduced.

Star–delta starting

The three sets of stator windings have end connections which are brought out to a starter box. Changeover contacts in the starter enable the six ends to be star-connected for

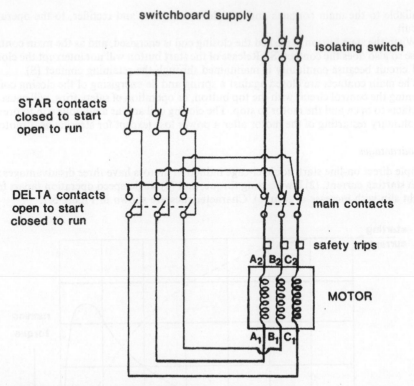

Figure 5.8 *Star–delta starter (see also Figure 5.4)*

starting (Figure 5.8) and then, as the rotor comes up to speed, to be reconnected in delta.

Star starting has the effect of reducing the voltage per phase to 57.7% of the line voltage. Starting current and torque are only a third of what they would be with direct on-line starting. The low-current start is obtained at the expense of torque and star–delta motors can only be used with light starting loads.

Automatic switching to the delta running condition is preferred to manual changeover which may be made too soon or too slowly and cause a current surge. In the delta running condition, phase voltage is equal to line voltage and the motor behaves as a straightforward squirrel-cage type.

Built-in interlocks or double-throw switches prevent star and delta contacts from being closed together. The starter is also designed so that star contacts have to be made before it is possible to change to the run position.

Auto-transformer starting

Conventional transformers have primary and secondary windings, but a single winding can be used as both primary and secondary for changing voltage. Primary voltage of 440 V applied across a single coil (Figure 5.9) can be tapped at some point along its length and at the end to give a secondary voltage with a value in proportion to the position of the tap.

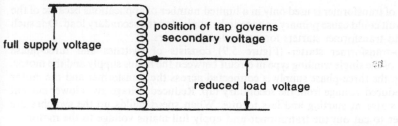

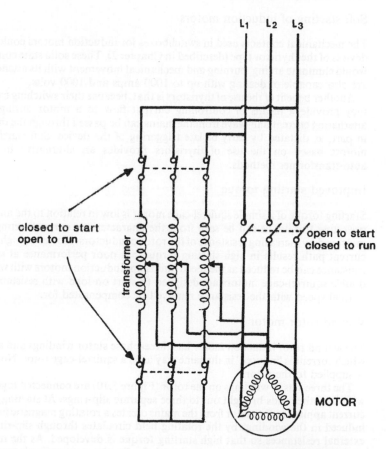

Figure 5.9 *Autotransformer starter*

This type of transformer is used only in a limited number of applications because of the risk that a fault could cause primary voltage to be applied to the secondary load. One such use is in auto-transformer starters for squirrel-cage motors.

The auto-transformer starter (Figure 5.9) consists of switches and three-phase transformer of the single winding type in circuit between the mains supply and the motor. For starting, the three-phase supply is connected across the transformer and the motor receives a reduced voltage from the secondary tap. Reduced voltage gives lower current flow in the stator at starting and less torque. When speed builds up the switches are changed over to cut out the transformer and apply full mains voltage to the motor.

Starting voltage can be varied by changing the winding tap; i.e. stator voltage and torque are adjustable.

Soft starting of induction motors

The mechanical contacts used in switchboxes for induction motors could be replaced by devices of the thyristor type (described in Chapter 2). These solid state controlled switches would eliminate arcing, burning and mechanical movement with its attendant wear. They are also capable of dealing with up to 1000 amps and 1000 volts.

Another benefit of the use of thyristors is that, because their switching can be controlled, they provide a means of controlling current flow to a motor during starting. Each alternating current half-wave from the mains can be passed through the thyristor in full or in part, as dictated by early or late triggering of the device. Soft starting of induction motors based on the use of thyristors provides an alternative to star–delta and auto-transformer methods.

Improved starting torque

Starting torque of a simple squirrel-cage motor is low in relation to the maximum possible operating torque, as can be seen from the characteristic curve. Starting torque can be improved by increasing resistance of the rotor conductors. However, high resistance in the current path results in high starting torque but poor performance at speed, unless the resistance can be reduced as the speed builds up. Induction motors with wound rotors and double squirrel-cage motors are designed to start on load with resistance but to run at normal speed with the resistance removed or compensated for.

Wound rotor motor

The wound rotor induction motor has three-phase stator windings and a wound rotor in which current is 'induced' in the same way as in a squirrel-cage rotor. No external current is supplied to the rotor.

The three sets of windings on the rotor (Figure 5.10) are connected together at one end, with the other ends brought out to three separate slip-rings. At starting, the three-phase current applied to the stator from the mains creates a rotating magnetic field. Current flow induced in the windings by the rotating field circulates through slip-rings, brushes and external resistances so that high starting torque is developed. As the motor runs up to speed the resistances are decreased and finally the current is short-circuited by the common connecting wire. The rotor is wound only so that it can be connected in series with external resistances to give high starting torque. After the starting period, induced current flows around the rotor windings in the same way as it circulates through bar conductors and end-rings of the squirrel-cage rotor. Interlocks ensure correct starting sequence.

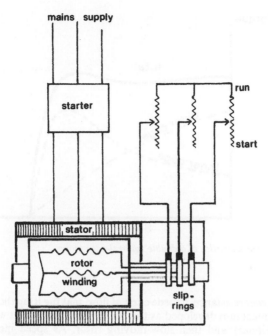

Figure 5.10 *Wound rotor motor*

Double cage rotor

The other method of obtaining high torque for starting on-load requires that the motor be fitted with a double cage rotor (Figure 5.11). The inner bars are of large cross section, low resistance copper, but bars in the outer cage are of small size and have higher specific resistance. The direct on-line method is used for starting and the three-phase supply to the stator produces a rotating magnetic field which cuts both sets of conductor bars. Initially current flow is induced mainly in the high-resistance outer cage. The inner conductor bars,

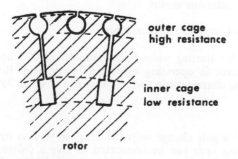

Figure 5.11 *Double squirrel-cage rotor*

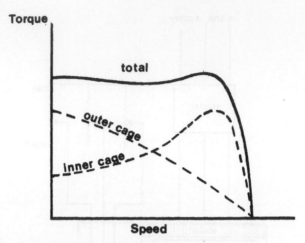

Figure 5.12 *Torque characteristics of a double squirrel-cage motor*

although having lower resistance, impede current flow, making it lag the rotating field; this lag is due to high reactance developed as the result of the fast rate at which the magnetic field cuts the stationary and then slow-moving rotor. As speed increases and slip is reduced, the rate of cutting drops and reactance of the inner cage lessens. The inner cage then becomes an easier current path and takes over as the means of producing torque (Figure 5.12). The characteristics of both high and low resistance cages combine to give good starting and running torque.

The direct on-line double cage motor provides a method of starting on-load without the windings, slip-rings and resistors of the wound rotor machine.

Speed variation

Single-speed operation of induction motors is suitable for most applications. However there are forms of speed control available and these include the reintroduction of resistance in wound rotor motors, and the changing of the effective number of poles which can be used with any induction motor.

Resistance control of wound rotor motors

The resistances used for starting wound rotor motors can be used to reduce speed, provided that the motor is operating against a heavy load. Reintroduction of the resistances changes the characteristic and allows a slip of up to 25%.

Pole change motor

The stator winding of a pole change motor is designed so that by means of switching arrangements the stator coils can be connected to give a different arrangement and effective number of poles. The arrangement shown (Figure 5.13) could be used to give two

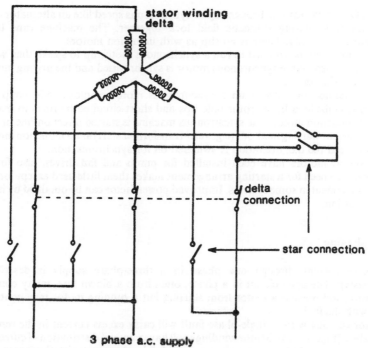

Figure 5.13 *Pole change arrangement*

fixed speeds with constant torque. In the low-speed version the stator poles are in series but delta connected; for high speed, a star configuration of poles in parallel is used.

For low-speed running five contactors are open and the three joined by the dotted line are closed. The three-phase supply through these poles is delivered to the stator windings in series in the delta arrangement.

In the high-speed version the stator windings are in parallel within a star arrangement. For this the three contacts joined by the dotted line are open and the other five closed (the two on the right connect the star mid-point and current is supplied to the star tips through the three contacts at the bottom of the sketch).

Synchronous motor

Rotors in synchronous motors are supplied directly with current from an outside source, unlike squirrel-cage and wound type rotors in which current is induced by the stator magnetic field.

Synchronous motors are constructed in the same way as alternators, with three-phase stator windings and rotors with salient poles. Like alternators, they require a d.c. supply to the rotor poles from an exciter via slip-rings. They are not self-starting. Connection of a three-phase input to the stator will produce a rotating magnetic field, but the effect on the poles is of equal attraction and repulsion and rotor inertia will prevent movement in either direction. Only if the rotor is brought up to synchronous speed like an alternator can the rotor poles and rotating magnetic field lock together. The machine runs then at synchronous speed only. There is no slip as with induction motors.

A pony motor can be coupled to run a synchronous motor up to speed, then with the excitation switched on the synchronous motor is synchronised and the driving induction (pony) motor shut down.

Another method of running up a synchronous motor employs solid copper bars permanently embedded in the rotor pole tips and short-circuited by rings to make it a temporary induction motor. The synchronous machine is started direct on-line or by one of the low-current starting methods (e.g. auto-transformer start) as an induction motor. At maximum speed the d.c. excitation is switched on for synchronisation.

Synchronous motors have been installed for pump and fan drives, also for main propulsion. The need for a starting arrangement makes them little used except for power factor improvement in some systems. Improved power factor can be obtained by increase of d.c. excitation.

Single-phasing

The loss of current through one phase in a three-phase supply is described as single-phasing. The open circuit in a phase, often from a blown fuse, faulty contact or broken wire, will prevent a motor from starting but a running motor may continue to operate with the fault.

A motor running with a single-phase fault will carry excess current in the remaining supply cables (Figure 5.14). Motor windings will have unequal distribution of current. The single winding A could have more than normal full load current with the motor on half load. Maximum motor loading could bring current in A to double or treble this figure.

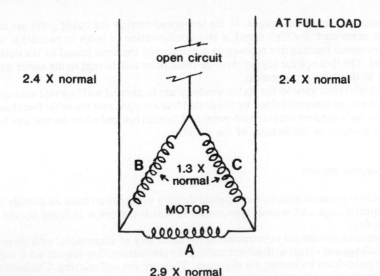

AT FULL LOAD

open circuit

2.4 X normal   2.4 X normal

B   1.3 X   C
     normal
MOTOR

A

2.9 X normal

Figure 5.14 *Single-phasing in delta-wound motor*

Current in the remaining two supply cables tends to rise in almost the same proportion as that in A. Flow through windings B and C in series becomes higher than usual at full load.

Single-phasing in a running motor can be detected by overload devices in the supply lines or through the overheating.

A single-phased motor arranged for automatic starting will not self-start and if overloads fail to operate may remain 'on' with consequent overheating. Overheating in a stalled or running motor will cause burnout of the overloaded coil.

It is possible for single-phasing in a lightly loaded motor to remain undetected by electromagnetic trips on the supply lines which monitor only current. Improved protection is given by thermistors placed in the windings to measure thermal effects.

Motor protection

Protection of motors is required mainly to prevent overheating which can cause deterioration of winding insulation and burnout, if severe. Overheating as the result of overload, stalling, single-phasing or prolonged starting period can be detected by a rise in line current and by temperature change. Overheating as the result of high ambient temperature or poor cooling due to blocked air passages can only be detected by temperature rise within the windings.

Overload protection is required for all motors of more than 0.5 kW although different rules apply to steering gear motors and others essential to safety or propulsion. A conventional electromagnetic overload trip must have a time delay dashpot (similar to those for d.c. switchboards) to allow for high starting current in direct on-line started induction motors. Unfortunately, an electromagnetic overload trip can be reset quickly and a motor restarted repeatedly with the result of excessively high winding temperature, unless a temperature trip is also provided.

Each of the three supply phases of the motor (Figure 5.15) is fitted with an overload

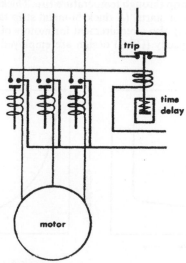

Figure 5.15 *Electro-magnetic overload device*

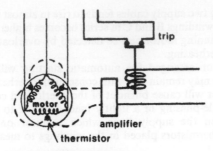

Figure 5.16 *Thermistor protection of motor windings*

relay. Such an arrangement should detect single-phasing, where the symptoms are of high current in two supply lines only as well as straight overload. Operation of any of the relays will close the circuit to energise the time-delayed trip.

The thermistor is a thermal device which can be used in conjunction with an electromagnetic overload trip. One of these inserted in each of the three windings (Figure 5.16) would detect overheating from any cause. Thermistors are available with either a positive or a negative characteristic (Figure 5.17). The former type are more definite in operation because there is a very sharp rise in resistance at a particular temperature (as opposed to gradual drop in resistance of the other sort). Positive thermistors can be connected simply in series and the very small current which passes through them normally is cut off by the effect of overheating in any one of them. Cessation of the minute checking current is used as the signal to operate the motor trip.

Alternative methods of detecting overload current employ directly or indirectly heated bi-metal strips (Figure 5.18). Excessive current in any of the supply cables will cause deflection of the bi-metal strip through temperature rise. Thickness of the strips is used to delay tripping when a motor starts. (A thick bi-metal strip takes longer to heat.)

Short-circuit protection is also a requirement for motors of over 0.5 kW. Fuses of the cartridge/high rupture capacity (HRC) design are employed to provide the necessary

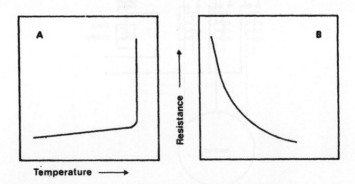

Figure 5.17 *Characteristics of (A) positive and (B) negative type thermistors*

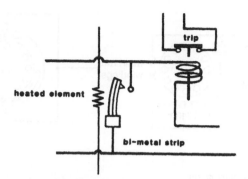

Figure 5.18 *Overload trip using indirectly heated bimetal strip*

rapid interruption of high fault current. Because short-circuit current may be high enough to damage normal motor contacts, the fuses may be arranged to break first in the event of short circuit. The secondary function of fuses is to provide back-up for the other protective devices.

Single-phase induction motors

Motors for washing machines and other domestic equipment operate from the single-phase low voltage supply. Motors in the machinery spaces are connected to the three-phase higher voltage supply. A three-phase supply to a delta or star winding produces a rotating magnetic field as described in the section on three-phase motors. A single-phase motor has two main windings in which the single-phase supply develops opposite polarities that alternate only. This effect is sufficient to keep the rotor turning once it has started, but an extra pair of windings (Figure 5.19) with a lagging or leading current is necessary to start the motor initially. The current from a single-phase supply can be caused to lag by incorporation of an induction coil or lead, with the use of a capacitor.

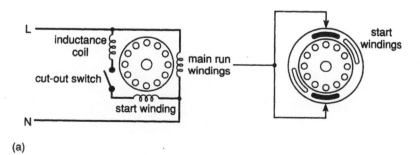

(a)

Figure 5.19a *Inductance coil starting for single-phase motor*

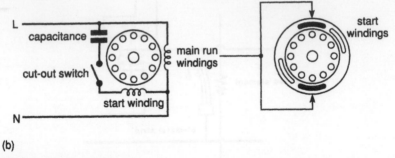

(b)

Figure 5.19b *Capacitor-start motor*

Induction coil

If an induction coil is used, it is connected in series with the extra starter windings and a cut-out switch. At starting, current from the mains passes straight to the main windings but via the coil to the starter windings. The a.c. supply voltage produces in the coil an increasing and then decreasing current, first in one direction and then in the other. The magnetising effect of the current flow builds up a magnetic field. As the field builds, it also cuts the coils and fulfils the generator rule. An e.m.f. that opposes supply voltage is generated in the coil and this slows down the rate at which supply current changes in the coil.

The effect is to delay current flow to the extra starter windings. A rotating magnetic field is therefore produced in the two sets of windings, which acts on the rotor in the same way as in a normal induction motor. The rotor turns and, as it runs up to speed, a device (centrifugal trip) cuts out the starting circuit with its induction coil and windings.

Capacitor starting uses a similar circuit.

CHAPTER SIX

D.C. Generators

Operation of a d.c. generator relies (as with alternators) on the principle that when magnetic lines of force are cut by a conductor (Figure 6.1) a voltage is induced in the conductor. Size of induced voltage and resulting current are dependent on magnetic field strength, length of conductor and speed of cutting.

The direction of current flow is dictated by the relationship between magnetic field and direction of movement of the conductor. It can be found from Fleming's Right Hand Rule, which is applied to give direction of conventional current flow during generation, also shown in Figure 6.1.

A simple generator can be constructed from a loop or coil of wire mounted on a spindle and arranged for rotation between opposite magnetic poles (Figure 6.2). The field-cutting

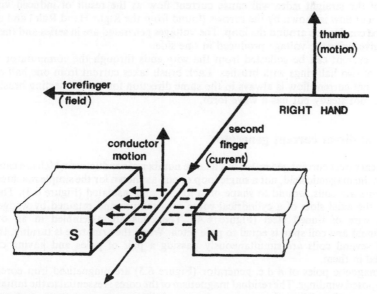

Figure 6.1 *Right Hand Rule of generation*

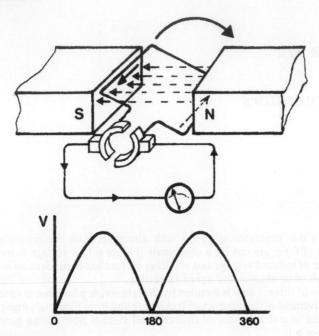

Figure 6.2 *Simple generator*

action of the straight sides will cause current flow as the result of induced voltage. Direction of flow is shown by the arrows (found from the Right Hand Rule) and can be seen to be continuous around the loop. The voltages generated are in series and therefore add to give twice the voltage produced in one side.

Direct current can be collected from the wire ends through the commutator which consists of two half-rings with brushes. Each brush takes current from one half-ring in turn, so that current flow is always in the same direction for each collecting brush. The output is not steady but has a wave form.

Practical direct current generator

A practical direct current generator has a large number of conductors which are caused to rotate in the magnetic field, not a single loop or coil as shown for the simple machine. The conductors are coils, wound to shape on a former and insulated (Figure 6.3). They are fixed in the axial slots of a cylindrical rotor or armature and retained by wedges and binding wire of tinned steel (Figure 6.4). The coils are assembled in an overlap arrangement and coil span is equal to pole pitch. When the armature is turning, the two sides of several coils are simultaneously passing a pair of poles and having current generated in them.

The magnetic poles of a d.c. generator (Figure 6.5) are magnetised iron cores with superimposed windings. The residual magnetism of the cores is essential to the initiation of current generation. When the generator is started, the rotating armature windings have

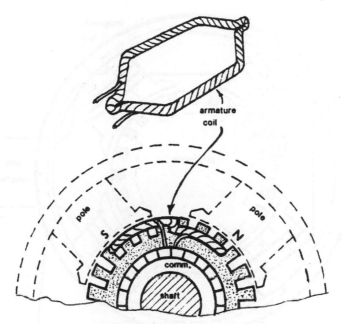

Figure 6.3 *Armature for a DC generator (or motor)*

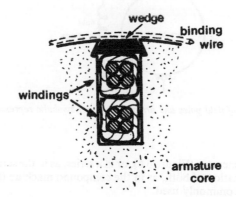

Figure 6.4 *Detail of winding in slot*

current induced in them only because they cut through the weak fields produced by residual magnetism in the iron cores of the poles.

The small generated current passes to the windings on the poles via the commutator and brushes. The electromagnetic effect greatly boosts the initially small residual field. Field pole windings as shown in Figure 6.5 can be connected in parallel with the armature and

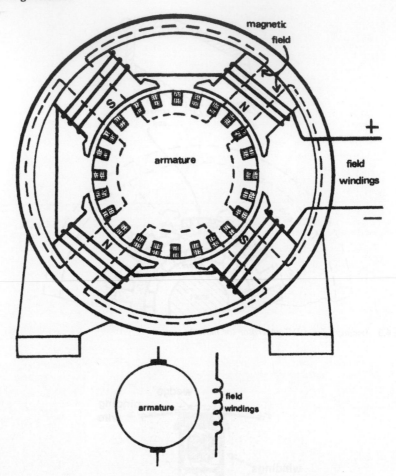

Figure 6.5 *Arrangement of field poles and armature with its symbolic representation*

load, as in the shunt generator (Figure 6.6), or in a series, as in the series generator (Figure 6.7), or with a combination of both, as in the compound machine (Figure 6.8). The last arrangement is most commonly used.

Shunt generator

The field pole windings of the shunt generator (Figure 6.6a) are permanently connected in parallel with the armature. Thus the small current produced in the armature conductors when they cut through the weak residual magnetic fields of the poles at start-up is delivered directly to the windings and serves to boost the magnetic field strength. Voltage builds up as the armature conductors cut through the stronger field, and in turn the higher voltage

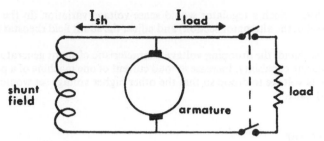

Figure 6.6a *Shunt generator*

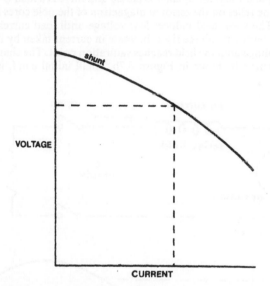

Figure 6.6b *Shunt generator characteristic voltage droop with load increase*

pushes a greater current through the field windings. The mutual build-up causes voltage to reach its peak very quickly.

Voltage is restricted to a certain maximum by the use of many turns of fine wire in the shunt windings which gives high resistance and limits current flow through them. Further restriction with control can be incorporated by fitting a variable resistor (or rheostat) on the shunt field circuit.

A shunt field generator with no load will run with its maximum voltage. If the output is used for load, the terminal voltage will droop (Figure 6.6b). Decrease of terminal voltage with steady increase of load current from zero can be plotted to show the typical characteristic for a shunt generator.

The drooping voltage characteristic of shunt generators makes them unsuitable for supplying current to equipment which requires constant voltage, unless a voltage regulator is fitted. Such a regulator would sense voltage variation (in the same way as those described in the alternator section) and adjust the shunt field rheostat to correct the deviation.

On the other hand, the drooping voltage characteristic of shunt generators gives them good load-sharing capability. Increase in load current of one machine of a pair in parallel would cause its voltage to droop so that the other higher voltage set would immediately take the load back.

Series generator

The field poles of a series generator are connected in series with the armature but the circuit is only complete when the breaker is closed and there is a load (Figure 6.7a). With no load, the machine relies on the residual magnetism of the pole cores for production of terminal voltage. The weak field induces low voltage until load current augments field strength, when the terminal voltage rises. Increase in current taken by the load results in greater terminal voltage until the field reaches saturation point. The characteristic voltage and load current curve is shown in Figure 6.7b. Small initial e.m.f. is due to residual magnetism.

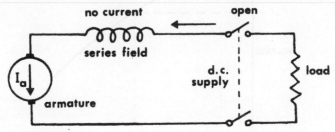

Figure 6.7a *Series generator*

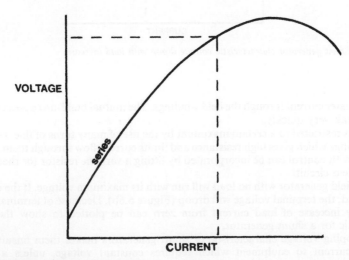

Figure 6.7b *Series generator characteristic*

Series windings are heavier than those for shunt fields, because they carry full load current and must have low resistance to keep volt drop small. Correct flux density is obtained with fewer turns of wire because of the higher current.

Series generators have limited use for special applications.

Compound generator

The different windings and characteristics of shunt and series generators are combined in the compound generator to give almost constant voltage over the operating range (Figure 6.8c). Both shunt and series windings are mounted on the same pole cores but one set is connected in parallel with the armature, the other in series.

Shunt windings consist of many turns of fine wire, as in the shunt generator. Heavy wire or strip used for the series windings is wrapped over the shunt coil and both sets are suitably insulated and protected.

The shunt field provides full voltage after the initial build-up, at no load. With increasing load the shunt field voltage droops but the series field voltage rises and an almost level characteristic is obtained. The shunt, series and compound curves are shown in Figure 6.8c. The ratio between shunt and series windings can be altered to give over-,

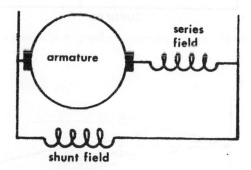

Figure 6.8a *Long shunt connected generator*

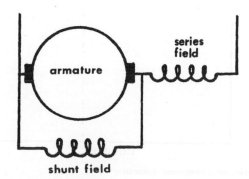

Figure 6.8b *Short shunt connected generator*

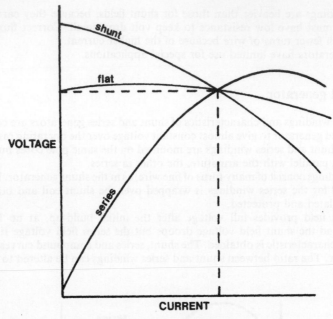

Figure 6.8c *Characteristic of compound generator, combining those of shunt and series types*

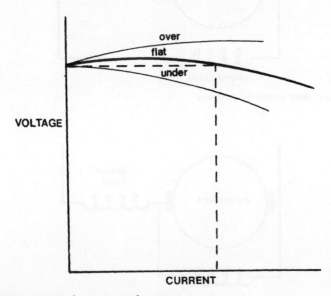

Figure 6.9 *Characteristics for a compound generator*

level- or under-compounding as shown in Figure 6.9. Slight under-compounding would give good load-sharing as obtained with the shunt machine.

Discrepancy in load between two under-compounded sets would leave the machine which had the greater share of load with a lower voltage. The generator with the higher voltage would tend to take more load until the voltages were equal. However, generators may be level or over-compounded so that equalising connections are used to stabilise load-sharing.

The basic connection of shunt and series fields can be by long or short shunts (Figure 6.8a,b). With the long shunt method, the shunt windings are in parallel with both armature and series field. A short shunt is connected in parallel with the armature only. A later diagram (Figure 6.16) shows a generator with shunt and series fields, shunt rheostat, interpoles and equalising connection.

Armature windings and the commutator

The simple commutator shown in Figure 6.2 is inadequate for a generator with a large number of overlapping windings which are simultaneously cutting through the fields of several pairs of poles. The practical commutator is made up of a large number of copper segments clamped to form a drum-like extension on the end of the armature (Figure 6.10). Each segment has connections from two adjacent conductor coils, as shown in the sketch of a lap-wound machine.

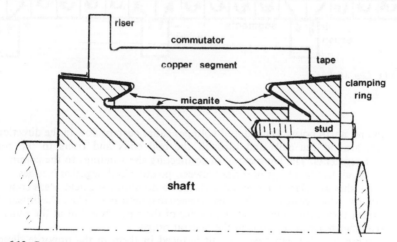

Figure 6.10 *Commutator segment assembly*

Lap-wound armature

The overlapping coils in Figure 6.11 are shown with one side as a full line and the other as a broken line to represent coil sides in tops and bottoms of slots respectively. Coils A, B, C, D and E are, as the armature rotates, simultaneously cutting the magnetic fields associated

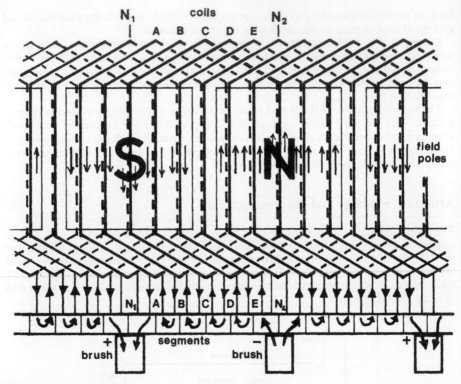

Figure 6.11 *Lap winding*

with south and north poles. The induced voltages cause current flow in the directions of the arrows, around the coils and through the coil ends and risers to the copper commutator segments. The lap method of connecting the windings to the commutator segments puts all five windings in series between positive and negative brushes.

Carbon brushes are placed in contact with the commutator segment areas coinciding with the range at which coils are in the gaps between magnetic pole fields. The brushes take part in current flow with windings at both sides of their position, but at the same time momentarily short-circuit the coils in the neutral range.

Coils moving past the gap have current induced in them in the opposite direction because of the change of magnetic field.

Overlapped coils are drawn separated for clarity in Figure 6.12.

Brushes

Brushes are housed in non-ferrous holders (Figure 6.13) in which they are free to slide. Contact with the commutator is maintained by spring-loading suitable to the brush material. Holders for a number of brushes are attached to arms and are adjustable. Sets of

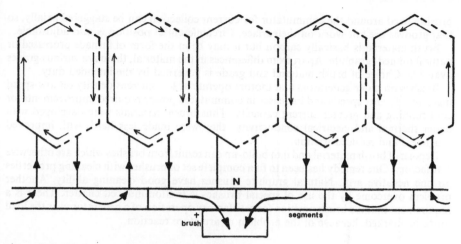

Figure 6.12 *Windings separated to show current flow*

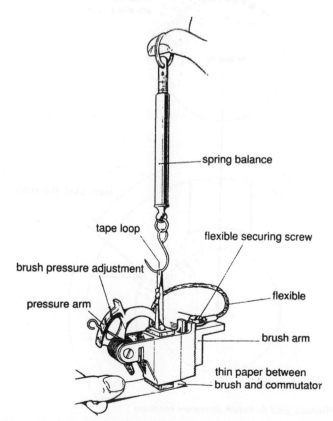

Figure 6.13 *Brush and holder showing method of checking pressure (courtesy of NEI-APE-W.H. Allen)*

brushes fixed around the commutator for current collection can be staggered axially, so that grooves are not worn on the surface. Circumferential position is also adjustable.

Brush material is basically carbon but it may be in the form of furnace processed or natural (mined) graphite. Apart from differences in the material, there are various grades available. Choice of brush material and grade is governed by the intended duty.

Brushes on older generators and motors operate at low current density on low-speed machines. Higher speeds and increases in commutated power required improvements for cool running and greater current capacity. Thus brush materials were developed with better thermal and electrical conductivity: they were made less hard, with improved toughness and reduced friction.

Deposit of brush material and film build-up can result from brushes which are otherwise satisfactory. One remedy has been to fit a complete set of brushes with cleaning properties on one negative arm. Natural graphite brushes have good cleaning ability. Another problem overcome by the use of special brushes is friction-induced chatter. There is a requirement that the brush position for satisfactory operation of a generator over the load range be marked, because of the problem of armature reaction.

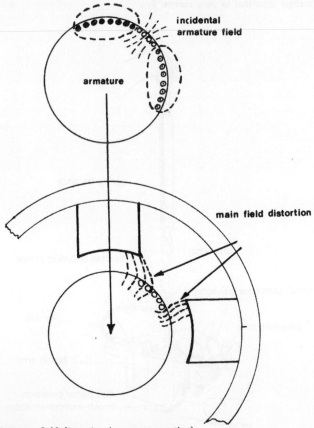

Figure 6.14 *Magnetic field distortion (armature reaction)*

Armature reaction

Ideally, the magnetic field being cut by the armature conductor ends sharply at the edge of the poles and current flow in the armature conductors would also cease at the gap before starting again in the opposite direction. In practice, the main magnetic field is distorted by an incidental magnetic field resulting from current flow in the armature windings (Figure 6.14). This 'armature reaction' means that current continues to be generated as coil sides pass into the physical gap between poles, and unless brushes are set further around on the commutator they would be short-circuiting coils in which current was still being generated. The effect would be to cause arcing between brushes and the commutator segments.

Interpoles and compensating windings

Unfortunately, the main flux distortion is greater with a strong incidental armature field and less when it is weak, so that changes of load current would require constant adjustment of brushes to a different neutral position. Automatic brush adjustment is possible, but the neutral plane can be prevented from shifting by arranging an opposing effect controlled by changes in armature current. Interpoles and compensating windings connected in series with armature windings provide the solution.

Interpoles or commutating poles

The polarity of each interpole in a generator (Figure 6.15) is opposite to that of the preceding main field pole and the same as that of the following pole, in the direction of rotation. The interpoles are in series with the armature and, because they carry full load current, have heavy windings. Interpoles are represented in diagrams by the symbol used in Figure 6.16. Flux provided by the few turns of interpole winding opposes the tendency for the field of the preceding opposite pole to be dragged around with the armature, so reducing the amount of armature reaction. Increases in armature current and field strength which would further distort the main magnetic field are counteracted by a similar increase in interpole flux, so that change of armature reaction is cancelled by the changing flux of the interpole. Over the load range of a generator the effect of the interpoles is to prevent change of armature reaction so that, with a stabilised neutral position, there is no need for constant brush adjustment. However, brush position is always a little to one side of the mid-point. Shaft magnetisation is avoided by arranging that current flows in opposite directions through main pole and interpole circuits. The currents and magnetic effects then tend to neutralise each other. This prevents generation of strong currents that could promote bearing damage.

Compensating windings

These are also connected in series with the armature so that current flow through them is the same as, and varies with, that in the armature windings. They are fitted in slots in the pole faces parallel with the armature windings, but current flows through them in the opposite direction so that they cancel the change in armature reaction caused by variations in armature current.

Compensating windings are used in machines having large load swings and consequently massive changes in armature reaction such that rapid movement of the flux induces high

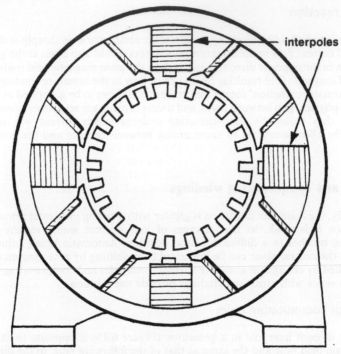

Figure 6.15 *Position of interpoles between main poles*

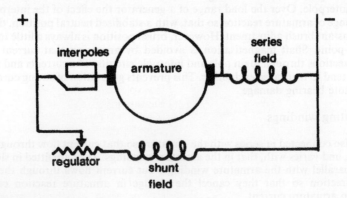

Figure 6.16 *Connection of interpoles in a d.c. generator*

voltage in the armature windings. Without compensation, a sudden load increase could cause a voltage rise in the armature windings sufficient to produce flashover between commutator segments.

Operation

Paralleling of a d.c. generator

The machine to be connected to the switchboard in parallel with another must, after initial starting and a check of satisfactory running, have its voltage brought up to 220 volts. The rheostat on the shunt field circuit is used for this adjustment. The voltage of the running machine is corrected if necessary and then the breaker of the incoming machine is closed. Finally, both generator rheostats are used to bring up the load current contribution of the newly started machine and to reduce that of the other, slowly, until suitable sharing is achieved.

Machine fails to excite

Sometimes when a generator has been started, the voltmeter fails to register a reading in the normal way. The pointer, instead of swinging up a little and then definitely moving along the scale as the rheostat is turned, may remain at the zero position. It might show a small negative reading. The most likely causes for this behaviour are the loss or reversal of the residual magnetism of the poles, respectively.

The residual magnetism of the poles is essential for the initial generation of the current necessary for the further build-up of shunt field strength. Correct direction of residual magnetic field is also vital.

Both of these common faults are remedied by passing a current through the shunt field coils, in the direction which will remagnetise the iron core in the right way. Current for restoration can be obtained from another d.c. generator or from a battery. The operation is carried out with the faulty machine stopped.

Using another generator

The brushes of the faulty generator are taken out of contact with the commutator segments either by inserting a sheet of insulating material (thick chart paper) beneath them or by lifting them out of the holders. If the latter, they are rested on the holders or allowed to hang on the connecting wires. The rheostat is turned to zero so that there is full resistance to current flow through the shunt field. The breaker for the machine which has failed to excite is closed while holding off the low-volts trip to allow this.

Closure of the breaker permits current from the running machine to pass to the faulty machine through the bus-bars. Backward flow of current will not pass through the armature or the series field, because of the lifted brushes. Such flow through the low-resistance armature and series field would cause damage and, being in the wrong direction in the series field, would completely reverse the pole residual magnetism. Current flow in the shunt field will be in the correct direction to re-establish residual magnetism. Current is in any case limited by the high resistance of the shunt winding (many turns of fine wire) but is controlled by the rheostat. This is turned to the low-resistance position, so that normal current passes through the shunt winding and the field is restored. The rheostat is then turned back to full resistance before opening the breaker, so the breaker does not arc due to interrupting a current flow nor due to the stored energy in the coils.

After releasing the breaker, the brushes are replaced and the generator is restarted, when residual magnetism will initiate voltage generation; this should build up in the normal way.

Using a battery

Residual field of the pole cores can be restored with a 12 V battery which is connected exclusively across the shunt field with the machine stopped. Current flow in the right direction for a few seconds only will establish the field.

CHAPTER SEVEN

D.C. Switchboards and Distribution Systems

The switchgear of d.c. switchboards, unlike that of enclosed a.c. equipment, was mounted on what was literally a board or screen of panels. On the front the open main circuit breakers, distribution breakers, knife switches and instruments were easily accessible for inspection and maintenance. Safeguards against contact were only an insulated handrail and a rubber mat. A closed gate normally prevented entry to the passage behind the board where the copper bus-bars and rheostats were mounted.

Enclosed or dead-front boards are only required for d.c. installations if the voltage is greater than 250 volts. Most of the d.c. switchboards on British vessels just prior to the general change to a.c. were for 220 volt supplies.

Main circuit breaker

The brush-type moving contacts of the generator breaker (Figure 7.1) are closed against a resisting spring by a lift-and-latch handle. The handle, when raised, slips the operating pin over the latch, where it is clipped into position. Downward movement of the handle pushes the cam plate against the hinged contact holder. The leverage and pressure applied force the brush contacts to spread on the fixed studs. This brushing action is beneficial in that contacts are wiped and any film or deposit is removed. Vibration with consequent burning or welding is prevented by the pressure, sustained by the resilience of copper strips, which are clamped in the manner of a leaf spring. Conductivity of the copper is slightly reduced by a small amount of alloying material added to prevent work-hardening and to give the necessary resilience.

The tripping of a main breaker due to overload or when load current was being carried would cause arcing and damage to the main contacts. Arcing contacts are therefore fitted and arranged to make first and break last, to protect main contacts from burning. They are of metallised carbon, sintered silver–tungsten or other material which does not weld at high temperature; such compositions are not suitable for carrying high current over long periods.

There is a set of main and arcing contacts for each connection to be made from the generator to the switchboard busbars. They are fixed together by bars for simultaneous movement by one handle. Thus, for a two-wire insulated d.c. system the triple-pole circuit breaker connects the positive, negative and equalising connections in one operation. The

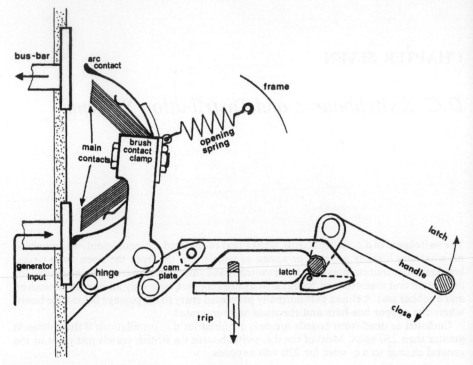

Figure 7.1 *Main circuit breaker for d.c. switchboard*

equalising contact is given a lead so that it closes first. Short-circuit between poles is prevented by insulation.

The breaker is held in the closed position by the latching device but is easily opened by the trip. Tripping is automatic in the event of short-circuit, overload, reverse current or low-voltage faults. For normal opening, the trip can be operated manually after reducing the load with the rheostat.

Generator protection

Generator circuit breakers are fitted with a number of protective devices. In general, for machines rated at over 50 kW and intended for parallel operation, these are:

(1) an instantaneous short-circuit trip,
(2) an overload trip with up to 15 seconds' time delay,
(3) preference trips for shedding non-essential load,
(4) a reverse-current trip,
(5) an instantaneous under-voltage release (machine not generating)
(6) an under-voltage trip with time delay.

On the distribution side there are protective trips and fuses which will disconnect motors or other circuits in the event of a local fault. They also give protection to generators and, with their shorter time delays and lower current settings, will normally clear a fault before the main breaker trips. The term 'discrimination' is used to describe the grading of protective device time delays and current settings throughout a distribution system.

Short-circuit, overload and preference trips are described below. They are all activated by excess current and can be based on a coil connected in series to carry load current.

Instantaneous short-circuit trip

A simple electromagnetic relay (Figure 7.2) will operate almost instantaneously when the magnetic effect is increased sufficiently by excess current to attract the iron 'armature' against the resistance of the spring. The moving arm can open the breaker through a trip circuit or by the direct movement of a mechanical linkage.

The generator may be capable of supplying a short-circuit current of up to twice the full load current. Damage is prevented by setting the trip at much less than this figure.

Overload trip

Various types of time delay are used for overload trips to ensure that the generator breaker will not be opened due to a momentary current surge, but only if the excess current persists. Overheating and damage of machine windings and cable results from prolonged overload current above the maximum rating.

Time lags of the dashpot type (Figure 7.3) delay tripping for up to a maximum of 15 seconds when excess current attracts the plunger to the solenoid. The piston can only move up as the silicone fluid is displaced from top to bottom through a small hole in the

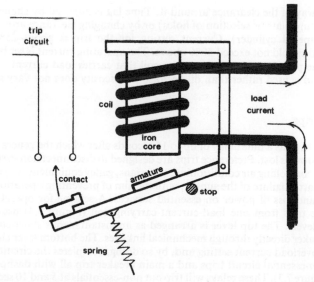

Figure 7.2 *Instantaneous short-circuit trip*

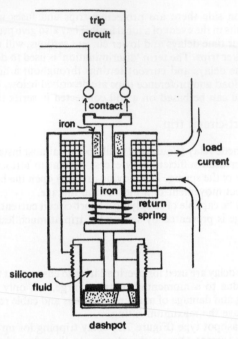

Figure 7.3 Overload trip

piston or by way of the clearance around it. Time lag is adjusted by changing hole size (different floating plate or selection of holes) or by changing clearance around the piston (adjustable tapered cylinder). Current setting for the trip is about 25% above the maximum and should not exceed 50% above. The operating current is set by varying the position of the plunger with respect to the coil that carries load current.

Silicone fluid is used rather than oil because its viscosity does not vary so much with temperature.

Preference trips

The overload trip has a time delay of up to 15 seconds after which the generator breaker is opened and power is lost. Preference trips are designed to disconnect non-essential circuits (e.g. breakers controlling air conditioning, some fans, galley equipment etc.) in the event of overload or partial failure of the supply, with the aim of preventing operation of the main breaker trip and loss of power on essential services. A scheme for operating all of the overload-type trips from one load-current-carrying coil (Figure 7.4) uses two instantaneous trip levers. The top lever is arranged as an instantaneous short-circuit trip and opens the breaker directly through mechanical linkages. The bottom lever closes instantly at the lower overload current setting and, by so doing, completes the circuit through two (or more) non-essential circuit trips and a main breaker trip all with dashpot time delay (similar to Figure 7.3). These relays will trip out non-essentials at 5 and 10 second intervals and finally, if the overload persists, the main breaker after 15 seconds. Warning of

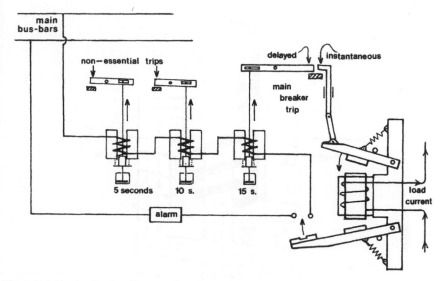

Figure 7.4 *Overload and preference trips*

overload is given by the alarm. Overload protection is provided on both poles (see Figure 7.7).

Reverse current trip

Loss of excitation or prime mover power causes a drop in generator voltage. If running in parallel, normal full voltage from the other generator(s) will cause current to be fed back into the faulty machine. Construction of d.c. generators and motors being similar, the reverse current will motor the generator with the possibility of damage to its prime mover and overload of the remaining power source. With direct current systems a reversal of flow can be monitored by using the change of direction of the associated magnetic field in a coil (carrying load current) or in the cable itself. Reverse current devices are required to be fitted in the pole opposite to that in which the series windings are connected when there is an equaliser (i.e. the positive pole).

The armature spindle of the device shown (Figure 7.5) carries the trip which opens the breaker. Anticlockwise movement of the armature operates it. During normal working load, current passing through the winding on the middle leg of the E-shaped laminated iron core creates a magnetic field. The direction of this field and of those on the outer two legs is as marked on the sketch. The high-resistance (voltage) coils on the outer legs are fixed across the main positive and negative generator outputs (see Figure 7.7).

The armature is pivoted on the load current coil leg, and the load current magnetic field extends through the iron core and the armature. This flux (shown by dotted lines) is superimposed on those due to the voltage coils (shown by the chain-dotted lines). It increases field strength on the left-hand side, pulling the iron armature clockwise against the stop, and reduces that in the right-hand side.

Reverse current in the centre load current coil leg reverses its magnetic field; the

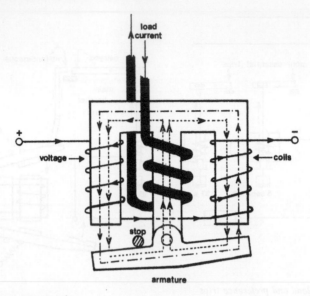

Figure 7.5 *Reverse current trip*

superimposed field now weakens the left-hand field and, in strengthening the flux in the right-hand gap, pulls the armature anticlockwise to trip the breaker.

Under-voltage release/trip

Means to ensure that the main circuit breaker cannot be closed unless the generator is running and generating correctly can be incorporated into the under-voltage trip. The latter is intended to open the breaker at loss of excitation for any reason and has a time delay.

The solenoid (Figure 7.6) is wired across the main leads on the generator side of the circuit breaker and, when not energised, leaves the plunger in the trip position and the breaker cannot be closed. Normal voltage output with the machine running energises the coil and the plunger is pulled down against the loading spring to release the trip. This allows closure of the breaker and means that the device is set.

Parallel operation of d.c. generators (with equaliser)

The voltage of a d.c. **shunt** generator drops as its load current increases (see Figure 6.8). This characteristic makes shunt machines ideal for stable parallel operation because any increase of the load taken by one generator will reduce the voltage of that set, so that the other, higher-voltage generator will tend to take the load back. A machine taking less than its proper share of the load will generate higher voltage and take load from the other.

Compound d.c. generators operating in parallel may be unstable because they lack the drooping voltage characteristic. They are usually over-compounded, which gives them a rising voltage with increase of load current. A machine taking more than its correct share

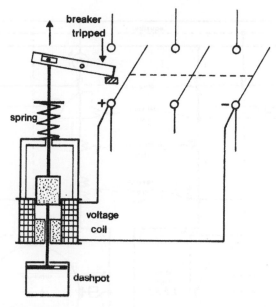

Figure 7.6 *Under-voltage release*

of the load would therefore generate higher voltage than before and actually increase its load further.

Even identical machines have slight differences in speed or other characteristics which will cause one to develop slightly greater voltage than the other and to increase its load, although to begin with each has an equal share. The increased load current in one machine, in passing through the series field of that machine, increases the field strength to give it higher voltage and load. Lower load current of the other machine weakens its series field and reduces its voltage and load. Without some outside correcting factor the load swing will continue until one generator has the full load and the other has none. To help stabilise the parallel operation of over-compounded and level-compounded (which have a part-rising voltage characteristic) d.c. generators, an equaliser is incorporated.

The equaliser is a low-resistance circuit made between paralleled compound generators via a special bar in the switchboard. It is shown by the dotted line in Figure 7.7, connecting the armature ends of the series coils of the two generators. If the no. 2 generator develops a slightly higher voltage and starts to take more load, the greater current is not confined to the series coil of machine no. 2. The equaliser passes part of the extra current to the series coil of machine no. 1. Field strengths of both machines are kept equal and a massive load swing is prevented.

During the operation in parallel of compound-wound d.c. generators, there is a continual hunting backward and forward of load as first one and then the other machine generates slightly higher voltage. The equaliser limits the change of load by paralleling the series fields, so ensuring equal current and field strengths.

Equaliser contacts in the main breaker are arranged to close before and open after the main contacts.

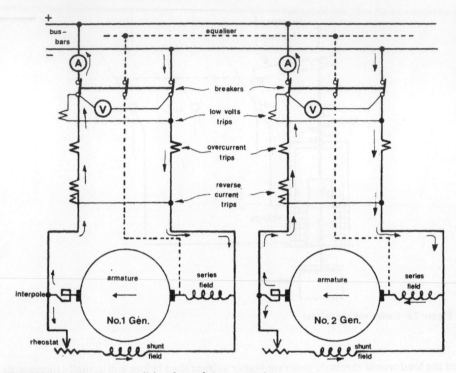

Figure 7.7 *Generators in parallel with equalising connection*

Prime movers

Load-sharing between d.c. generators is also stabilised if the prime movers, through their governors, are given a slightly dropping characteristic. This means that as load increases, the governor undercorrects by a very small amount, and that over the load range from zero to full load there is a speed loss of perhaps 5–6%. A machine taking more than its share of load would run slower and the faster generator (with higher voltage from the speed) would take it back.

Mechanical governors have a naturally drooping characteristic.

Earth lamps

Normal lights when supplied from a d.c. system can be connected across the full 220 volts between positive and negative leads. Earth lamps are small 30 watt lamps suitable for use with a 220 V system, but being a pair in series (Figure 7.8) the voltage effect is lowered and the lights burn with reduced brightness. The mid-point is permanently earthed by being attached to the ship's steel structure.

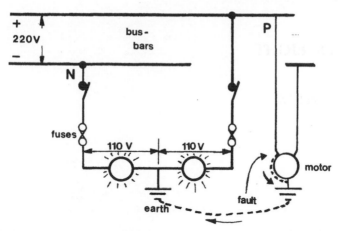

Figure 7.8 *Earth lamps for a d.c. switchboard*

The main insulated two-wire d.c. distribution system is completely isolated electrically from the ship's structure (unlike the formerly used single-wire system with hull return), except for the mid-point earth between the earth lamps. A leakage of current through faulty insulation is made obvious by its effect on the lights because the fault provides an alternative current path, bypassing the earth lamp on the fault side. A severe fault will provide a path with much less resistance for current than the one through the earth lamp: the lamp will lose brightness or go out.

A typical failure is shown in the sketch, with insulation breakdown allowing a motor casing to be 'live' and the fault current travelling to the ship's structure via the safety earth connection on the motor casing. The lamp on the fault side (in the example, the positive) darkens but the other one brightens. The brighter light is thus because it is connected on its own across the full 220 V supply, albeit through an earth fault.

Another aspect is that with good insulation the positive bus-bar has a potential of + 110 V relative to the ship's structure and the negative − 110 V, due to the mid-point earth which must be at the same potential as the hull. A fault on one side puts that side to the same potential as the hull and the other side 220 V different. The lamp connected between the fault side and the mid-point earth has very little or no potential across it and goes out. The lamp connected across the sound side and earth has the full 220 V and is therefore brighter than before.

The importance of earth lamps is that they give immediate indication of a current leakage that is potentially dangerous. A number of fires have been the result of ignition by the spark at an earth fault in wiring. The most dangerous condition occurs when there is a fault on both positive and negative sides of the distribution at the same time. This amounts to a short-circuit across a 220 V supply, with the degree of current flow depending on the severity of the faults. An additional problem is that equal-size faults will affect the lamps to the same extent and slight dimness of both lamps may not be noticed. The check for such a double problem is to switch off one lamp and look for an increase in brightness of the other.

CHAPTER EIGHT

D.C. Motors

Motor construction and operation

A d.c. motor is constructed in the same way as a d.c. generator, with inward projecting field poles in the yoke and a rotating armature with conductors in the slots (Figure 8.1). Input current is applied to the shunt and/or series windings on the field poles and also, through brushes and commutator, to the lap or wave wound armature conductors which are momentarily passing under the poles. Current in the pole windings produces fixed magnetic fields. Current in the armature conductors sets up a small magnetic field in each of those under the poles. Reaction of the fields associated with the conductors and the magnetic fields of the poles produces a turning effect in accordance with the Motor Rule. This rule applies to all motors and can be summarised with reference to the simple sketch.

Current passing through a single conductor (Figure 8.2) sets up a circular flux and if the conductor is positioned in and at right angles to another field it will tend to be moved out of the field (as arrowed). Direction of movement of the conductor is dictated by the magnetic field and current flow in the conductor. Mutual directions can be worked out using Fleming's Left Hand Motor Rule. With forefinger and second finger aligned with the field and conventional currentflow respectively, the thumb points in the direction of relative motion of the conductor. Field direction is from north to south and conventional current flow is from the positive supply terminal and around the circuit to the negative supply terminal.

The large number of conductors on a d.c. motor armature are connected in a lap or similar winding as in a d.c. generator. Current applied through the brushes and commutator flows through conductors adjacent to the poles in such a way as to give an overall turning effect in the same direction. For clockwise rotation of the motor shown (Figure 8.1), input current would flow into the page in the conductors under south poles and out of the page in those under the north poles.

Resistance starting of d.c. motors

Current flow from the mains supply through the armature conductors of a d.c. motor fulfils the motor requirement and causes rotation. However, rotating of the conductors through the steady magnetic field of the poles also fulfils the generator condition. A voltage termed **back e.m.f.** (electromotive force) is induced in the armature conductors.

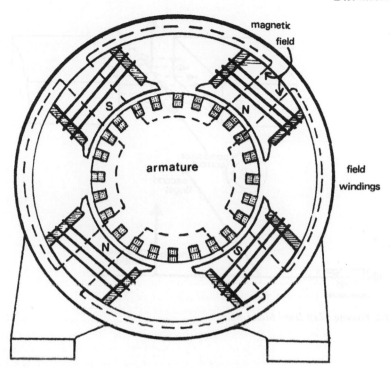

Figure 8.1 *D.c. motor with shunt and series (compound) windings*

The back e.m.f. is in opposition to and a little less than supply voltage. The net effect, with supply voltage being greater, is that current flows into the low-resistance armature as the result of the small difference between the two voltages.

Back e.m.f. is not developed unless a motor is running, and the application of full supply voltage without an opposing voltage being generated in the motor would result in massive current flow through the armature. To protect the armature during the run-up from zero to full speed, the starter has a resistance in series with the armature (Figure 8.3) to restrict current. As motor speed and back e.m.f. build up, the resistance is cut out by movement of the starting handle.

Shunt motor starter

The starter shown in Figure 8.3 for a shunt motor illustrates the basic arrangement with the return spring, hold-on (no volts) coil, and overload trip protective devices. The contact arm is in the off position and for starting is moved to the first stud contact with the isolating switch closed. This closes the circuit so that current flows through the starting resistances and armature (being restricted by the resistances). It also takes a path from the first stud through the shunt field via the hold-on coil. Application of current to the field poles sets up a magnetic flux which rotates the armature because of the current in the

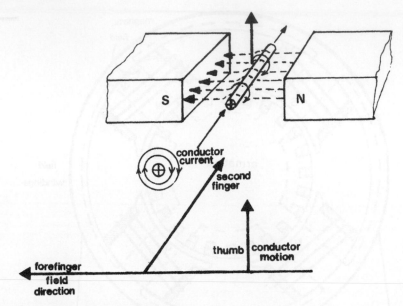

Figure 8.2 *Fleming's Left Hand Motor Rule*

armature conductors, and in turn a back e.m.f. is generated in the windings as they cut the main magnetic field. This counter-voltage opposes the supply voltage and reduces its effectiveness in pushing current through the armature.

As the motor speeds up, armature current drops until the contact arm is moved on to the next stud, so cutting out some of the resistance and again increasing the current. A pattern of current increase and reduction is produced as the contact arm is moved across and resistance cut out. At the final stud, the arm is held against the pull of the return spring by the hold-on coil. In this position full mains voltage is available to the armature, but back e.m.f. limits the effective voltage and current flow. The starting operation takes a few seconds, during which the contact arm is moved slowly but continuously from left to right in the starter shown. Overheating of the resistance results if the arm is left at an intermediate position.

Shunt field current is taken from the first resistance stud, so that at the initiation of the starting sequence full current passes through the shunt field. As the contact arm is moved to cut out resistance in the path for current flow to the armature during motor run-up, the same resistances are left in the path for current flow through the contact arm to the shunt field. The slight reduction in the small shunt field current and magnetic strength has the effect of producing a moderate speed increase (see Field control, below).

Types of d.c. motor

The field poles of d.c. motors, like those of d.c. generators, can be connected with a shunt, series or compound arrangement and motors are classified by the type of connection.

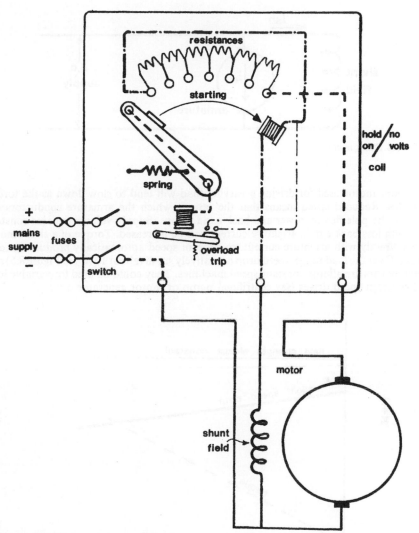

Figure 8.3 *Resistance starter for a shunt motor*

Some special-purpose motors are separately excited, and small motors may have permanent magnet poles.

Shunt motors

The pole windings of a simple shunt motor are connected in parallel with the armature (Figure 8.4). Current input to the armature will vary but the field poles will take almost constant current and as a result produce an almost unchanging magnetic field.

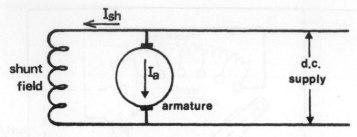

Figure 8.4 *Shunt motor*

A shunt motor used for driving a varying load will tend to slow down as the torque increases. Reduced speed means that the rate at which the armature conductors cut through the pole flux is slower and back e.m.f. drops. Supply voltage remains constant and, with less back e.m.f., armature current will be increased. Torque of a shunt motor varies directly with armature current, so drop in speed automatically produces greater torque. Over the load range, speed drop is relatively small (about 10%, see Figure 8.5) and shunt motors are almost constant-speed machines. They could be used for variable load and constant-speed drives (say centrifugal pump or motor generator drives).

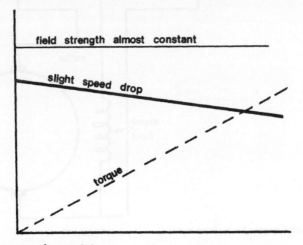

Figure 8.5 *Shunt motor characteristic*

Series motors

A simple series motor (Figure 8.6) has field pole windings which are connected in series with the armature. Load current flowing consecutively through field and armature is therefore the same in each. Strength of the field will vary with armature current.

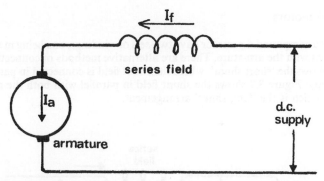

Figure 8.6 *Simple series motor*

Armature current is governed by the difference between supply voltage and back e.m.f. A drop in back e.m.f. causes a larger margin of effective voltage and this in turn produces greater current flow. Back e.m.f. in a series motor changes as speed varies with load, and also because the magnetic field strength changes with load current. The speed torque curve for a series motor (Figure 8.7) shows that, with low load, speed is excessive but the motor has very high starting torque. With high load, speed is low.

The series motor's characteristic of high starting torque would make it suitable for deck machinery, but the dangerously high light-load speed has to be limited by a small shunt winding in addition to the series field. A simple series motor requires a safety speed cut-out or permanently coupled load to prevent it from running away at loss of load.

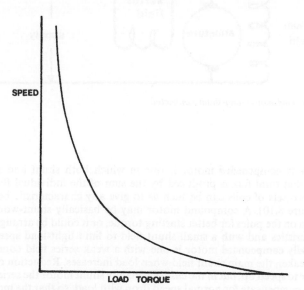

Figure 8.7 *Series motor characteristic*

Compound motors

The poles of a compound d.c. motor have two sets of windings, one being in shunt and the other in series with the armature. There are alternative methods of connecting the fields. Figure 8.8 shows the 'short shunt', where the shunt field is connected in parallel with the armature only; Figure 8.9 shows the shunt field in parallel with both the armature and series field, which is the 'long shunt' arrangement.

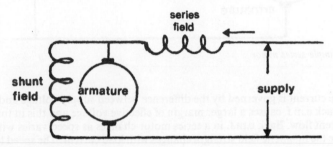

Figure 8.8 *Compound motor short-shunt connected*

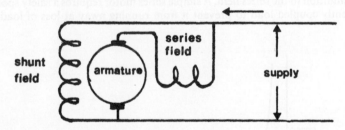

Figure 8.9 *Compound motor long-shunt connected*

A **cumulatively compounded** motor is one in which both shunt and series coils are connected so that total flux is produced by the sum of the individual fluxes. The ratio between the two sets of coils can be such as to give any characteristic between the two extremes (Figure 8.10). A compound motor may be basically shunt-wound with a few series windings on the poles for better starting torque, or it could be arranged with mainly series characteristics and with a small shunt field to limit light-load speed.

A **differentially compounded** motor is one with a weak series field connected so as to oppose and weaken the main shunt field when load increases. Reduction of field strength causes a motor to speed up, but in the case of a mainly shunt motor the series windings can be designed to compensate for normal speed drop with load, so that the motor has almost constant speed over the load range. Differential motors can be unstable.

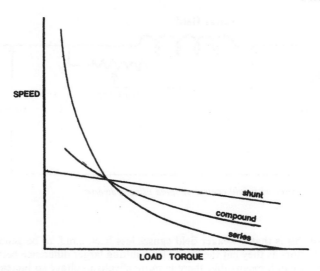

Figure 8.10 *Compounding characteristics*

Speed control of d.c. motors

The effect of changing load on the simple motors previously described is that some degree of speed variation occurs. Controlled speed change is imposed by using means to alter current through the field windings or armature. Conventional methods which employed variable resistances or diverters have been replaced in modern d.c. motors by thyristor armature control. The Ward-Leonard system is still used.

Field control

Change of current through field pole windings is easy to achieve by use of a rheostat (variable resistance) in series with a shunt motor field (Figure 8.11) or a diverter for the field of a series motor (Figure 8.12). The weakening of the magnetic fields of the poles by reducing field current has the inverse effect of causing the motor to speed up. The

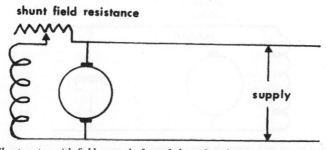

Figure 8.11 *Shunt motor with field control of speed through a rheostat*

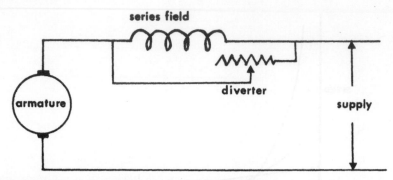

Figure 8.12 *Series motor with field control of speed using a diverter*

explanation for this is that a weaker field causes less back e.m.f. to be generated in the armature conductors as they cut the flux. The resulting larger difference between supply voltage and back e.m.f. means that there is more effective voltage to increase armature current. This method is useful for changing speed at the top end of the range, but it is not possible to reduce speed below that which is obtained when full shunt-field current flows.

One limitation is that weakening of the magnetic fields of the poles permits them to be more easily distorted by the rotating armature (armature reaction). Greater advantage can be taken of field control if the machine has interpoles to limit the armature reaction.

Armature control

This can be used to give a speed range from zero to full with speed approximately proportional to armature voltage.

Rheostat controllers are variable resistors connected in series with the armature (Figure 8.13) to alter the effectiveness of supply voltage across the armature. Speed changes as a consequence. The method was used for d.c. deck machinery out of necessity but it was wasteful of energy. Apart from its inefficiency, there was the problem of resistances getting hot due to the proportion of large armature current passing through them. The speed range extends from zero to full.

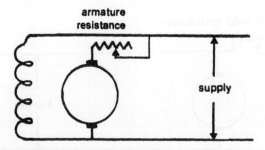

Figure 8.13 *Shunt motor with armature speed control*

Thyristor control applied to the armature of a d.c. motor uses a rectifier arrangement that includes silicon-controlled rectifiers. In the example (Figure 8.14) the armature current is taken through a three-phase rectifier which has an SCR and an ordinary rectifier on each section. Triggering by a gate signal (Chapter 2) which is early or late gives control. Current to the field pole windings is taken through a simple three-phase rectifier, and remains constant.

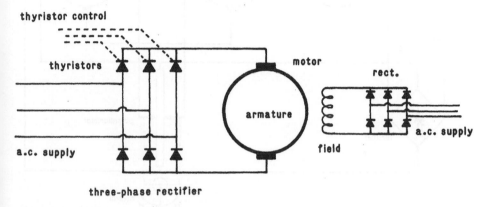

Figure 8.14 *Thyristor control of d.c. motor speed*

Ward-Leonard system

This system, used for fine control of d.c. electric motor speed from zero to full in either direction, is also able to give the motor a robust torque characteristic. The system was used for the motors of electric (as opposed to hydraulic) steering gears of ships with d.c. electrical power, and it is used today on ships with a.c. electrical power for deck machinery such as the windlass.

The working motor (Figure 8.15) which powers the steering gear, windlass or other equipment is a d.c. machine, because speed control of these is easy. The method is to alter the voltage applied through the brushes to the armature windings of the d.c. motor; no change is made to the current in the field windings. The voltage is increased or decreased, not with the use of resistances but by arranging an individual d.c. generator with controllable output voltage as the power supply for the armature of the working motor. Speed and direction of the working motor vary with the magnitude and direction of applied voltage. Current for the windings of the field poles is derived from the source of main power. Where there is a.c. main power, current for the windings is transformed to lower voltage and rectified.

The d.c. generator is driven by a simple one-speed, direct on-line start, squirrel-cage motor (a.c.-powered ship). Output of the generator is varied by changing the current to its field windings through a variable control resistor (potentiometer). As magnetic field strength is altered by the change of field current, so too is the generated voltage. Switch of direction of current flow through the field poles, also with the potentiometer, will cause the

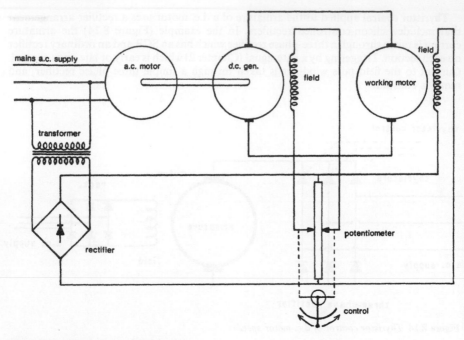

Figure 8.15 *Ward-Leonard system*

direction of the pole magnetic fields to change. This changes the direction of generated current supplied to the motor and thereby also the running direction of the motor.

The control lever can, by moving the potentiometer contacts in opposite directions away from the mid-position shown in the sketch, set the polarity and strength of the generator field poles. By governing generator output to the armature of the working motor, this in turn gives stepless speed control of the working motor, in either direction.

The system as used for steering gear operation on d.c. ships is complicated by the feedback which automatically brings the potentiometer to its mid/neutral position, when the rudder gear reaches the desired position.

Maintenance of d.c. motors

The d.c. motor has the same construction and therefore the same general maintenance requirements as the d.c. generator. Thus regular cleaning is needed to keep the air passages clear and to remove deposits of dust, oil and grease which could reduce the effectiveness of the insulation. Readings are taken of insulation resistance at intervals. The commutator and brushes are inspected and maintained as necessary to prevent sparking. Mechanical checks include bearings, holding-down bolts, and drive couplings or belts and pulleys. Motor starters and controllers require regular checks and maintenance of contacts, resistances and connections.

Before undertaking any work on electrical equipment, it must be isolated from the electrical supply completely and a notice should be fixed on the isolator. Personnel should be informed verbally and all circuit fuses removed and retained.

Contacts deteriorate in service as the result of burning and wear so that repair or renewal becomes necessary. Damage can be rectified by dressing with a fine file to restore the profile accurately across the surface. Emery is used for the final polish. Contacts become thinner as the result of wear and cleaning, so that there is loss of contact pressure. Loading is measured with a spring balance and corrected unless no adjustment is provided when contacts are renewed. Insufficient contact pressure results in welding together of surfaces and failure to open under fault conditions.

A thin smear of vaseline or other contact lubricant is applied to some types of contact after servicing. Light oil is used on pivot pins.

Resistors in motor controllers can be checked with a meter. Those of the cast grid type, supported at the ends, tend to crack due to vibration. Wound resistors fail through overheating or general deterioration of the insulation. Resistor boxes must be well ventilated when motors are in use. Connections are examined from time to time for corrosion or for overheating due to slackness. Where necessary they are disconnected, cleaned up and remade.

CHAPTER NINE

Safe Electrical Equipment for Hazardous Areas

Electrical equipment and flammable atmospheres

Hydrocarbon gases or vapours from crude oil form highly flammable mixtures with air (Figure 9.1) when present in the proportion of between 1% and 10% hydrocarbon with 99% down to 90% normal air. Below the lower explosive limit (LEL) the mixture is too lean to burn rapidly, although a lean mixture will burn slowly in the presence of a naked flame or a spark, as is proved by the operation of explosimeters in this range. Over-rich mixtures exist when the level of the hydrocarbon exceeds 10%.

The risk of explosion or fire within the cargo tanks, pumprooms and other enclosed areas of an oil tanker is high as a consequence of the very small quantity of hydrocarbon gas or vapour that makes up an explosive mixture. Within tanks, vapour is readily evolved from cargo: immediately above the surface of the liquid there tends to be a layer of undiluted hydrocarbon vapour. Further above the surface of the liquid, dilution by air (if present) increases towards the top of the tank, to give a mixture with a decreasing hydrocarbon vapour content. At the top of the tank, the larger quantity of air is likely to produce a lean mixture. Thus the free space within tanks above any liquid petroleum cargo, or in an empty tank from which the cargo has been discharged, is likely to contain a middle flammable layer sandwiched between lower over-rich and upper lean layers. Leakage of crude oil or products from pump glands would produce the same effect in pumprooms.

Inert gas is delivered to the cargo tanks of crude oil tankers to maintain a slightly higher than atmospheric pressure and so exclude air during operations. The inert gas can also be used to displace any air present in a tank after inspection or other period when the IG system has not been in use.

The IG system cannot be used for pumprooms or other spaces in which crude oil vapours can collect and therefore form a flammable mixture with air. In these areas and on deck, precautions to avoid the possibility of ignition are necessary.

There are no special measures taken in ordinary electrical equipment to encapsulate contacts which arc as they close or open. Thus light switches, torches, sockets and plugs, pushbutton contacts, relays, electric bells, starters, circuit breakers and open fuses are all potential sources of ignition should there be flammable gas, vapour or dust in the atmosphere. The danger from arcing contacts in a starter box or switch is not obvious because the arc is hidden. Sparks from an electric motor, particularly of the commutator

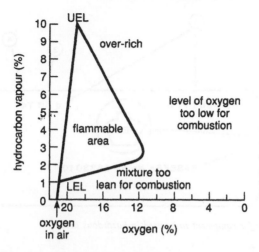

Figure 9.1 *Flammability of air and oil vapour mixtures*

type, the momentary glow of a broken light bulb filament, and arcing from a broken or damaged power cable are more apparent as sources of ignition.

Conventional equipment and cable are suitable for areas that are considered safe. Special regulations apply to hazardous zones requiring only safe equipment or none at all. An unusual condition (as with a gas leak in the home) can make a safe area potentially dangerous so that all machinery and equipment must be shut down.

Safe electrical equipment

The information needed when selecting suitable safe apparatus for use in hazardous areas is shown below as on a nameplate (Figure 9.2). The data consists of letter codes and references, which in the example apply to flameproof equipment. Other types of protection also use the marking for gas grouping and temperature.

The standard to which the safe apparatus is made may be British Standard BS 5501 or an equivalent from the IEC (International Electrotechnical Commission), or another standard. For flameproof equipment, details of construction method can be found in Part ¡5 of BS 5501. This item is indicated on the nameplate by (1) together with the code letter for flameproof apparatus (d).

The stylised crown with inscribed letters **Ex** is the symbol of the testing and certifying authority and this, indicated here by (2), is the British Approvals Service for Electrical Equipment in Flammable Atmospheres (BASEEFA). There are others. The 'Ex' represents the general term **explosion proof**, which is used in Europe for the safe types of electrical equipment.

Reference (3) shows the IEC grouping of apparatus with respect to certain gases. The roman numeral I is used for underground mining work, where methane and coal dust constitute the risks. Roman numeral II is used for equipment in industries other than mining. A sub-grouping for gas types is shown by a letter after the numeral. Thus group

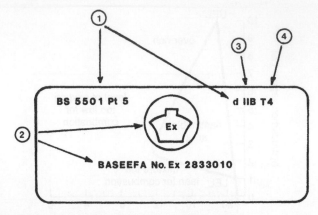

Figure 9.2 *Nameplate for equipment to be used in hazardous areas*

IIC is typified by the gas hydrogen, which requires an ignition energy of 29 microjoules. Group IIB is typified by ethylene, which requires an ignition energy of 69 microjoules. Group IIA is typified by propane which requires a higher ignition energy of 180 microjoules. This last gas type is obviously the most difficult to ignite. Further details of the gas classification can be found from the IEC publication or from BS 5345 Pt 1 (1976). Some countries have maintained their own grouping systems, which are slightly different and could be confusing.

The energy for ignition can be derived from a spark or from contact with a hot surface. Protection against sparks is necessary whenever flammable gas and air mixtures are present. Hot surfaces, however, will not cause combustion unless their temperature reaches the ignition temperature of the gas. (**Ignition temperature** is higher than **flashpoint**. A spark can cause combustion at flashpoint, but a hot surface will only cause combustion at the higher ignition temperature.) Maximum surface temperature classification is given by the T symbol, reference (4). The table gives the relationship between class number and the maximum surface temperature.

Class	Maximum surface temp. (°C)
T_1	450
T_2	300
T_3	200
T_4	135
T_5	100
T_6	85

Higher surface temperatures are shown as actual figures. Corrections are necessary if ambient temperature exceeds 40 °C or if equipment is in contact with other hot surfaces.

Equipment for hazardous areas is selected so that its maximum surface working temperature is less than the ignition temperature of any gas or vapour likely to be found in its hazard environment. Details of gas ignition temperatures and the T classifications can also be found in BS 5345 Pt 1 (1976) *Basic Requirements*, which is the UK code of practice for safe electrical installation in hazardous areas.

Flameproof (Ex d) equipment

If electrical equipment could be made with perfectly gastight casings, the possibility of gas seeping in and being ignited by a spark would not be a problem. Unfortunately, seals on the rotating shafts of motors or on the operating rods of solenoids are not gastight. Hermetic sealing of covers is not guaranteed: gaskets or jointing can deteriorate and face-to-face joints depend on quality of the machined surfaces and tightness of fastenings. Inaccurate threads on screwed cover joints or conduit can allow leakage of gas; so too can deteriorating cement or sealant on inspection windows or insulators.

The designers of flameproof equipment recognise that there is a risk of ignition and explosion of gas within some enclosures and counter this by making the casings strong enough to withstand such an explosion. Also, potential flame paths to the outside of the casing through joints or seals are made so as to extinguish any flame. Normally this is done in joints by making them face-to-face with a limited gap (Figure 9.3). Note that the possible flame path between the surfaces is made long enough to prevent flame emission (lengths for different types of joints are given in the construction standards), and packing or jointing is not used in the actual flameproof joint. The flame path through shaft seals is elongated by using a labyrinth design. Special consideration is given to cover fastenings which must be strong enough to withstand internal explosion and, as far as possible, gastight. Screws of special matterial must have a unique design so that ordinary screws cannot be used to replace them. Their heads are patterned for special keys. Screw holes must not pierce the casing.

The glass of inspection windows must be held against the casing by an internal clamping ring. Flame paths are incorporated and sealing cement must be durable.

Flameproof enclosures ('explosion proof' is the term applied in the USA) are used for equipment where sparking or arcing occurs during normal operation, as in a switch or starter. The spark is contained and likewise any flame or explosion (of gas which might enter the casing), preventing ignition of a surrounding explosive atmosphere.

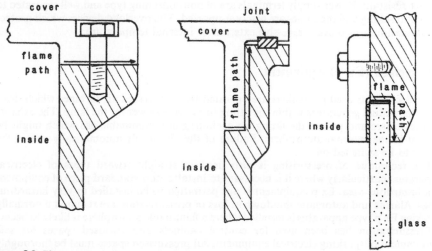

Figure 9.3 *Flame paths in Ex d equipment*

The maintenance of light fittings, switchboxes and other equipment which is designed and type-tested as flameproof must be thorough to preserve the 'as tested' condition. Inspection may reveal impairment of structure which would permit gas to enter, and also allow flame from gas ignited inside to reach the surrounding atmosphere without being cooled or extinguished.

Casings or terminal boxes may be corroded, cracked, broken or not properly closed by reason of missing or slack screws or nuts. Glass may be cracked, broken or not adequately cement-sealed. Deterioration of conduit, cable and cable boxes, or glands may be evident; also rubbing contact between fans or couplings and guards. Deterioration, wear or damage must be made good by replacement or repair followed by correct assembly after cleaning of flanges. Feeler gauges are used to ascertain that there is no gap in flanges greater than the specified maximum (which may be 0.15 mm). Any weatherproofing must be fixed as specified; paint is applied with care (not incendive aluminium paint) to protect against corrosion without affecting the flameproofing.

Electrical checks are for earth continuity and the satisfactory operation of any protective overload devices which are important for preventing overheating. Lamp rating and type must be checked in flameproof light fittings.

Increased safety (Ex e) equipment

A squirrel-cage induction motor is a type of electrical apparatus which does not normally have any associated arcing or sparking during operation and where running temperature is not excessive. Such equipment can be made safe for operation in areas made hazardous by the likelihood of flammable vapour, by the use of increased safety techniques.

Insulation for windings and cables is of high quality and protection is given against ingress of water or solids which could cause insulation breakdown. Breakdown of insulation due to overheating from overload is prevented by overload devices, which are an essential part of the increased safety technique. Adequate clearance is given to the fan and rotor to avoid mechanical sparks from rubbing contact and the casing is made impact-resistant. Power supply terminals are of non-loosening type and well separated to prevent tracking and the cables are firmly supported. The overload devices which protect the insulation also prevent excessive external or internal temperature.

Pressurised (Ex p) equipment

Some deck lights used for tankers are operated by compressed air turbines which drive small individual generators within the fitting to provide power for the lamp. The exhaust air pressurises and purges the fitting, so excluding any flammable gas which might be present in the external atmosphere. Failure of the air supply automatically causes the power to be switched off.

The technique of pressurising is also used in straightforward types of electrical apparatus, particularly where it is necessary to install a non-standard piece of equipment in a hazardous area. Ex p equipment is not permitted to be installed in very hazardous areas. Alarms and automatic shutdown at loss of pressurisation are required if normally sparking Ex p type apparatus is installed where a flammable atmosphere is likely to occur.

Pressurisation has been used for control cabinets and enclosed spaces for safe containment of sparking electrical equipment. All pressurised spaces must be thoroughly purged before the equipment is switched on.

Non-incendive (Ex n or N) equipment

The difficulty of absolutely sealing casings against the slow inward seepage of hazardous vapour which is permanently present in the surrounding atmosphere has been mentioned in the section on flameproof equipment. In an area where gas might only be present for a limited length of time, if at all, the problem of accumulation of vapour within a casing is not so serious. Thus the Ex n protection that can be used with lights in such slightly hazardous areas, for example, calls for (1) a conventional casing seal, (2) a surface temperature limit, and (3) protection against the ignition by operational sparks of any vapour which reaches the inside of the enclosure. Spark risk is negated by an enclosed-arc spark break device, or by ensuring that the spark is of low energy, or by hermetic sealing of the spark location.

Other methods of protection

Equipment for normal power operation can also be protected by oil immersion (**Ex o**), powder filling (**Ex q**), or by a special method of protection (**Ex s**).

Intrinsic safety (Ex i)

The effect of an electric arc or spark on flammable vapour is that the heat increases the energy of the vapour and air locally so that particles are activated and made ready for chemical combination. The energisation reaches a sufficient pitch after a very small time delay so that combustion results.

A weak spark or arc may not have sufficient thermal energy to heat the local flammable mixture at a rate faster than that at which the energy is dissipated to the further surrounding atmosphere. Without the rise of energy, no combustion reaction will occur. A weak spark or arc with a slow rate of heating the mixture may not persist long enough to complete the ignition at the end of the delay period.

An intrinsically safe circuit is one that is designed for a power so low that any spark or thermal effect produced by it, whether there is a fault or not, is incapable of igniting the surrounding flammable gas or vapour. It follows that intrinsically safe equipment is used in such circuits and is designed on the same basis, i.e. of being unable to produce a spark with enough power to ignite the specific flammable vapour or gas involved. Intrinsic safety technique requires not only that a system is designed for operation with very low power, but also that it is made invulnerable to high external energies and other effects.

Intrinsically safe apparatus is currently made to two standards of safety. **Ex i(a)** is the symbol for the higher standard, which requires that safety is maintained with up to two faults. This type of equipment can be fitted in any hazardous area. The other standard is given the symbol **Ex i(b)** and apparatus made to this specification is safe with up to one fault. The Ex i(b) products are not used in the most hazardous areas. Manufacturers of intrinsically safe apparatus state that this method of protection is suitable for electrical supplies at less than 30 volts and 50 milliamps. It is used extensively for instrumentation and some control functions.

Care is exercised in design that capacitance and inductance within the electrical installation are kept to a minimum, to prevent storage of energy which in the event of a fault could generate an incendive spark. Ex i systems are isolated from other electrical supplies even to the extent that the cables are not permitted to be in the same trays as those

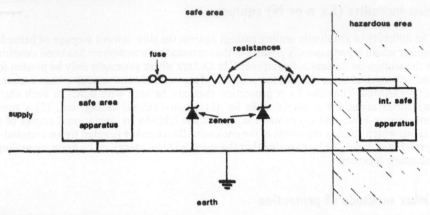

Figure 9.4 *Safety barrier for Ex i equipment*

of other cables (to prevent induction effects). Systems are earthed and protection is provided by inclusion of shunt diode safety barriers between hazardous and non-hazardous areas (Figure 9.4). The safety barriers have current-limiting resistors and voltage bypassing zener diodes to prevent excessive electrical energies from reaching the hazardous areas.

Neither certification nor marking is necessary if none of the following values are exceeded in a device: 1.2 V, 0.1 A, 20 microjoules, 25 milliwatts. However, great caution is needed when deciding whether apparatus will operate within all of these limits and any associated system would have to be certified as intrinsically safe.

Protection related to types of rotating machinery

With most types of electrical apparatus, a choice can be made as to the particular method of protection that is most suitable and economical. Four methods of explosion protection suitable for rotating machines are described below, together with any limitations. No electric motors or other rotating machines with these types of protection are permitted in pumprooms or other similar hazardous enclosed areas of ships, or in the most hazardous (zone 0: see page 129) areas of any onshore or offshore installation. Equipment with non-sparking Ex n/N protection can, in fact, only be positioned in zone 2 or least hazardous areas.

1. Flameproof (Ex d) protection

Flameproof Ex d protection can be applied to any type of rotating electrical machine but is intended for those where the ignition of a flammable atmosphere is likely because sparks or arcing occur during normal operation of the machine. Wound rotor induction motors with slip-rings and brushes, also commutator motors, are, because of sparking, likely to be protected by a flameproof (Ex d) enclosure. With such protection, a particular piece of equipment may be acceptable for zone 1 and zone 2 areas but the flameproof protection does not make it suitable for the most hazardous zone 0 locations.

Explosion test

The casing for flameproof protection of an electric motor or other type of rotating machinery must be strong enough to withstand pressure developed during an internal explosion. The strength of a prototype of the casing is put to the test by the sparkplug ignition of an explosive mixture containing a specified gas. The pressure developed by the explosion is found and may then be used as the basis for a further static or dynamic test at 1.5 times the explosion pressure or 3.5 bar, whichever is greater. For the protection to be deemed satisfactory, the casing should not have been damaged or deformed. (It is the casing or enclosure which is tested by the explosion and which must remain intact, but there is no stipulation about internal damage.)

Flameproof test

The requirement that the enclosure must also prevent transmission of any flame from an internal explosion to any explosive atmosphere surrounding the enclosure is tested by repeating the explosion test at least five times with the machine in a chamber which is filled with the same explosive mixture. If the mixture in the chamber is not ignited, then the second set of tests is considered to have been passed and a Flameproof Ex d certificate is granted. Recertification is needed for any modification of the casing.

Explosive mixtures vary in their ability to force flame through a gap of given dimensions; machines are grouped according to the design of joint (Figure 9.3) with respect to length of flame path, maximum gap and dimensions. Gases are listed for a given enclosure group.

General comments

A flameproof enclosure need not be hermetically sealed: emphasis is placed on design of potential flame paths for effectiveness in preventing passage of the flame, rather than providing a seal between any covers and the casing. With no jointing material permitted between cover and casing, the face-to-face gap, although limited in size, will allow the entry of small amounts of flammable gas or vapour to the enclosure. As in a Davey lamp, where a small quantity of gas burns in the presence of the flame, so in a flameproof enclosure for a machine with arcing contacts the gas will normally burn at a slow rate in the area where arcing occurs. If a larger quantity (within the explosive limits) accumulates, there may be a contained explosion.

There is no limit imposed by the flameproof (Ex d) requirements on the rating of a machine or its internal temperature other than what may be involved in limiting the external surface temperature of the enclosure or may be imposed by the rating of the insulation material.

2. Increased safety (Ex e) protection

Increased safety (Ex e) protection, as applied to an electric motor, is described in the section dealing with Ex e equipment (page 124).

3. Pressurised (Ex p) protection

Protection by pressurising is acceptable for all types of rotating machinery including motors with commutators or slip-rings. A requirement for protection by pressurising is that the casing is purged or scavenged so that any gas that may be present is completely displaced before the power is switched on. Air is normally used for pressurising but nitrogen or other inert gas may be employed.

The adequacy of pressurisation is tested by connecting a manometer (glass U-tube in which water is displaced by the internal pressure).

Efficiency of scavenging is also tested. The procedure involves displacing the atmosphere (air) within the enclosure with a safe, easily available test gas which has an appropriate vapour density and then, in turn, using the intended purging/pressurising gas to displace the test gas. A 'before and after' analysis of gas samples taken from a number of monitoring points is used to verify the purging.

This type of protection also involves a surface temperature classification.

4. Non-sparking (Ex N) protection

This type of protection can be used for squirrel-cage or brushless synchronous machines. The features are similar to those for Ex e protection but less demanding. The criteria for Ex N protection may well be met by standard machines. Certification is required, however, but the type of protection only makes the machine suitable for zone 2 areas.

Tanker installations

Regulations and practices applied to the installation of electrical equipment in tankers specify the types of safe equipment that can be fitted in the areas where flammable gas and air mixtures may be present. The degree of risk is not the same throughout the hazardous areas (Figure 9.5), which include cargo tanks and the spaces above them, pumprooms, cofferdams and any closed or semi-enclosed spaces with direct access to a dangerous zone.

Cargo tanks are permitted to have only intrinsically safe apparatus which is certified to the higher Ex i(a) standard.

Pumproom lighting must be flameproof (Ex d) and arranged with two separate and independent circuits. Intrinsically safe apparatus to Ex i(a) standard is also allowed.

Cofferdams adjacent to cargo tanks can be fitted with intrinsically safe equipment.

Deck areas above cargo tanks can be fitted with a wider variety of certified safe equipment including intrinsically safe apparatus and others such as flameproof (Ex d),

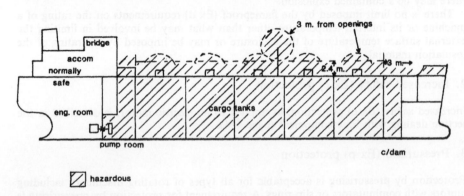

🔲 hazardous

Figure 9.5 *General tanker arrangement showing hazardous and normally safe areas*

increased safety (Ex e), and pressurised (Ex p), depending on distance from tank openings etc.

Forecastle spaces are considered hazardous if the entrance is within the area of deck which is hazardous and only certain safe types of equipment are permitted.

Selection of safe equipment for all tanker installations must take into consideration classification as to gas type (e.g. 11A – least incendive) and ignition temperature in relation to surface working temperature of the equipment.

Exceptionally, submerged electric motors are fitted in the cargo tanks of liquefied gas carriers. Provided that there is liquid in the tank, there will always be boil-off vapour at sufficient pressure to exclude air. Cargo pump motors are isolated when gas-freeing or at any time when there is a danger that air might enter the tank. Automatic shutdown and alarm in the event of low liquid level is required (initiated by low liquid level, low motor current or drop in pump discharge pressure).

Installations ashore

Codes of practice in the oil industry generally classify risk areas for the likelihood of flammable gas/air mixtures being present as:

Zone 0 – Explosive gas–air mixture continuously present or present for long periods. (Only intrinsically safe apparatus, certified to the higher Ex i(a) standard, is permitted where there is a continuous hazard.)

Zone 1 – Explosive gas–air mixture likely to occur during normal operation. (Equipment certified as intrinsically safe, Ex i(a) or (b); flameproof, Ex d; increased safety, Ex e; pressurised, Ex p; or having special protection, Ex s; is allowed in areas of intermittent hazard.)

Zone 2 – Explosive gas–air mixture not likely to occur and any occurrence would be for a short time. (Areas hazardous only under abnormal conditions are usually freely ventilated, and all of the safe types of equipment are permitted provided that they conform in other respects.)

There are variations in zone classification due to different codes of practice, methods of interpretation and number of zones (some countries recognise only two). The use of zone classification has been suggested for tankers.

Reference

Towle, L. C. (1985) 'Intrinsically safe installations on ships and offshore structures'. *Trans.I.Mar.E.* **98**, Paper 4.

CHAPTER TEN

Shaft-driven Generators

Auxiliary diesel-driven generators which run continuously for twenty-four hours a day both at sea and in port can be expensive in terms of the fuel cost and maintenance requirement. Maintenance is usually based on running hours which, with continuous operation, will add up to 720 per month or over 8000 in a year. Even where economy is achieved by the use of a blend of cheap residual with the more expensive distillate fuel, the accumulation of running hours still gives a maintenance requirement and it is possible that the workload will be heavier due to problems with the fuel.

A generator drive taken from the main propulsion system provides the means of reducing maintenance by avoiding the use of auxiliary diesel at sea. It also furnishes a method of obtaining electrical power from the cheapest fuel. Additional advantages brought by the installation of a shaft generator are that fewer diesel generator sets are needed and the shaft-driven machine can be of large enough capacity to take the full at-sea electrical load.

Power for the bow thrust while manoeuvring is provided exclusively by the main engine-driven alternator on some ships, with power for auxiliaries being provided at that time by two diesel sets. At sea, when the bow thrust is shut down and auxiliary load is transferred to the main engine-driven alternator, the diesel machines are stopped.

General arrangements

A generator positioned directly in the shaft line between the main engine and propeller (Figure 10.4) can be built so that its shaft is flange coupled as part of the intermediate shaft system, or the rotor can be based on a split hub which is clamped to a section of the main shaft. The outer part of the generator is supported on the tank top. The problem with this arrangement is that the air gap will vary due to hull flexure and weardown of bearings and an excessive clearance of perhaps 6 mm may be required.

Another arrangement, developed by the builders of large slow-speed diesels, has a large-diameter multi-pole short-length alternator (Figure 10.1) mounted on the forward end of the main engine. The projection adds only about one metre to the engine length and with the casing bolted to the engine, and rotor to the crankshaft, no extra support is needed and an excessive air gap is not required. With both schemes described above, the generator runs at what is likely to be a low engine speed.

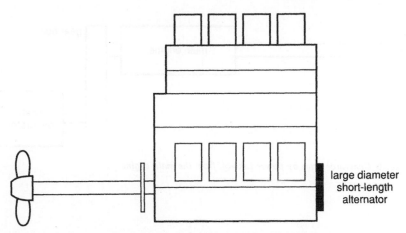

Figure 10.1 *Large slow-speed main engine with an alternator at the forward no-drive end*

Vee-belt drives have been used to step up d.c. generator speeds, but for alternators a step-up drive through gears is more usual. The drive can be taken from the intermediate shaft (Figure 10.2a) or from a main gearbox or from the engine itself (Figure 10.2b), to provide higher speed. The power take-off (PTO) from the engine through gears may be from the camshaft or crankshaft.

Another frequently used option (see Figure 10.5) to obtain higher speed is that of coupling generators to the non-drive end of the medium-speed main engines. The advantage of this arrangement is that the generator can be of large power because the medium-speed main engine is a prime mover that can be disconnected from the gearbox and used in port. Large-capacity cargo pumps for tankers or dredge pumps can be supplied with electrical power by such a system. The type of arrangement shown incorporates a high-voltage switchboard and distribution system which has been favoured for many installations.

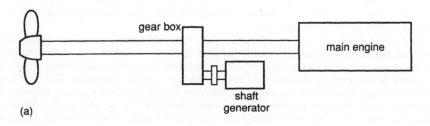

Figure 10.2a *Generator drive from the intermediate shaft*

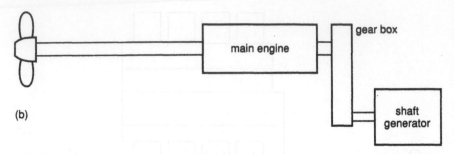

Figure 10.2b *Generator drive by power take-off from the main engine*

Speed variation

With d.c. systems the use of shaft generators was common because of the acceptability of limited speed variation. Fluctuations in speed of d.c. generators tend to cause change of voltage but this can be corrected by appropriate design of the excitation system and the incorporation of a voltage regulator.

The 60 Hz (or in some cases 50 Hz) frequency of an a.c. system is required to be maintained within very close limits; the speed of any main propulsion-driven alternator must therefore be kept within a very narrow band by the speed governor of its prime mover. One method for obtaining a.c. power from a shaft drive (Figure 10.3) utilised a belt-driven d.c. generator (with automatic voltage regulator) which supplied a d.c. motor, the latter being coupled to provide a mechanical drive to an alternator. The field of the shaft-driven d.c. generator is separately excited using power from the ship's three-phase a.c. supply. The excitation from the switchboard is delivered through a three-phase rectifier to the field as direct current. The AVR uses the output from the tacho-generator to monitor d.c. drive motor speed and thus the voltage output of the d.c. generator. The latter must be kept constant to maintain constant speed of the d.c. motor and the alternator that it drives.

The arrangement can accept and operate satisfactorily with a main engine speed variation of ±15%. When reducing main engine speed, the changeover from shaft generator to diesel drive may be automatic or manual.

The avoidance of speed deviation is achieved for many main propulsion-driven alternators by opting for a constant-speed engine with a controllable-pitch propeller. A suitable engine governor is essential.

Most ships have fixed-pitch propellers so that ship speed variation necessitates changing engine revolutions. To accommodate changes of engine and shaft alternator speed, a constant-speed power take-off may be installed or, more usually, output from the alternator is delivered to the electrical system through a static converter. The converter accepts a range of generated frequency but delivers a supply at the frequency required by the system. Static frequency converters have been developed for use with shaft alternators where the speed range extends from 40% to 100% of the rated speed of the main engine.

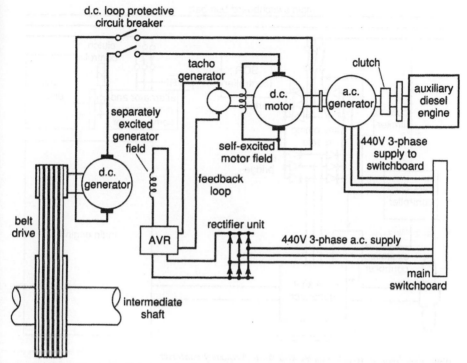

Figure 10.3 *Alternator driven by shaft-driven d.c. generator and motor*

Static frequency converter for a shaft generator

The converter system shown (Figure 10.4) serves the shaft generator of a ship with a fixed-pitch propeller and a large main-engine speed range. The shaft generator must supply full output over the permitted speed range, and to achieve this at the lower end (i.e. down to 40% of the rated speed), it is overrated for higher speeds.

The a.c. shaft generator itself is a synchronous machine which produces alternating current with a frequency that is dictated by variations in engine speed. At the full rated r.p.m., frequency may match that of the electrical system.

The output is delivered to the static converter, which has two main parts. The first is a rectifier bridge to change shaft generator output from alternating to direct current. The second part is an inverter to change the d.c. back to alternating current, at the correct frequency.

Alternating current from the shaft generator, when delivered to the three-phase rectifier bridge, passes through the diodes in the forward direction only, as a direct current (see Figure 2.12, page 25). The smoothing reactor reduces ripple. The original frequency (within the limits) is unimportant once the supply has been altered to d.c. by the rectifier.

The inverter for transposition of the temporary direct current back to alternating

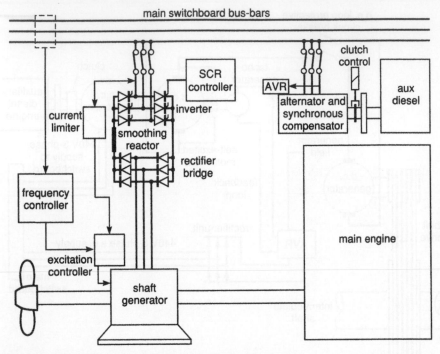

Figure 10.4 *Shaft generator system with static frequency converter*

current is a bridge made up of six thyristors. Direct current, available to the thyristor bridge, is blocked unless the thyristors are triggered or fired by gate signal. Gate signals are controlled to switch each thyristor on in sequence, to pass a pulse of current. The pattern of alternate current flow and break constitutes an approximation to a three-phase alternating current.

Voltage and frequency of the inverter supply to the a.c. system must be kept constant within limits. These characteristics are controlled for a normal alternator by the automatic voltage regulator and the governor of the prime mover, respectively. They could be controlled for the shaft alternator inverter by a separate diesel-driven synchronous alternator running in parallel. The extra alternator could also supply other effects necessary to the proper functioning of an inverter, but the objective of gaining fuel and maintenance economy with a shaft alternator would be lost. Fortunately the benefits can be obtained from a synchronous compensator (sometimes termed a synchronous condenser), which does not require a prime mover or driving motor except for starting. The compensator may be an exclusive device with its own starter motor or it may be an ordinary alternator with a clutch on the drive shaft from the prime mover.

The a.c. generator set that fulfils the role of synchronous compensator for the system shown (Figure 10.4) is at the top right of the sketch. The diesel prime mover for the compensator is started and used to bring it up to speed for connection to the switchboard. The excitation is then set to provide the reactive power, and finally the clutch is opened,

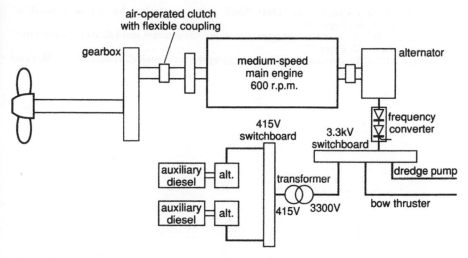

Figure 10.5 *Shaft generator supplying a high voltage system*

the diesel shut down and the synchronous machine then continues to rotate independently like a synchronous motor, at a speed corresponding to the frequency of the a.c. system.

A synchronous compensator is used with the monitoring and controlling system, to dictate or define the frequency. It also maintains constant a.c. system voltage, damps any harmonics and meets the reactive power requirements of the system and converter, as well as supplying, in the event of a short circuit, the current necessary to operate trips.

The cooling arrangements for static frequency converters include the provision of fans as well as the necessary heat sinks for thyristors.

High voltage system for a dredger

The necessity for high electrical power for dredge pumps, bow thrusters and other machinery has led to an increased investment in high voltage systems. The function of the main engine of many vessels is now divided between propulsion and that of acting as prime mover for generators with a large power output for auxiliary machinery.

The diagram of the main propulsion and electrical supply scheme for a dredger (Figure 10.5) illustrates the changing role of the main engine. In this arrangement, when the vessel is manoeuvring or dredging and using less propulsive power, the excess is made available to the bow thrust or dredge pump. Full engine power is used only when on passage.

References

Gundlach, H. (1972) 'Static frequency converters for shaft generator systems'. *Trans. I. Mar.E.*
Hensel, W. (1984) 'Energy saving in ships' power supplies'. *Trans.I.Mar.E.* **96**, Paper 49.
Mikkelsen, G. (1984) 'Auxiliary power generation in today's ships'. *Trans.I.Mar.E.* **96**, Paper 52.

Murrell, P. W., and Barclay, L. (1984) 'Shaft driven generators for marine application'. *Trans.I.Mar.E.* **96**, Paper 50.

Pringle, G. G. (1982) 'Economic power generation at sea: the constant speed shaft driven generator'. *Trans.I.Mar.E.* **94**, Paper 30.

Schneider, P. (1984) 'Production of auxiliary energy by the main engine'. *Trans.I.Mar.E.* **96**, Paper 51.

CHAPTER ELEVEN

Electric Propulsion

An electric propulsion arrangement for a ship is often simply described as a diesel electric or turbo-electric system. It is characterised only by the type of prime mover, with no reference to the type of electric propulsion motor, the generator or the electrical power system.

The electrical side of the system will be based on a direct current or an alternating current motor, coupled to the ship's propeller shaft, with speed and direction of propeller rotation being governed by electrical control of the motor itself or by alterations of the power supply. An electric motor used with a controllable-pitch propeller is arranged for either constant or variable speed operation.

For a direct current (d.c.) propulsion motor, the electrical power may be from one or more d.c. generators or it may be alternator derived and delivered through a rectifier as a d.c. supply. The rectification scheme can incorporate speed control and the means of reversing.

Power for an alternating current (a.c.) propulsion motor is supplied by an alternating current generator (alternator). The prime mover that provides the generator drive may be a diesel engine, a gas turbine or a boiler and steam turbine installation.

Electric propulsion has been used mainly for specialised vessels rather than for cargo ships in general. These include dredgers, tugs, trawlers, lighthouse tenders, cable ships, ice breakers, research ships, floating cranes and vessels for the offshore industry. The main advantage lies with the flexibility and absence of physical constraints on machinery layout. Support ships for the offshore industry, particularly those with two submerged hulls, can use electric propulsion motors to give high propulsive power in the restricted pontoon space, while generators and their prime movers are housed in the large platform machinery space. Electric power can be used for the self-positioning thrusters and other equipment, as indicated in Figure 11.7, as well as for main propulsion.

Passenger ships with electric propulsion benefit because the number of generators in operation can be matched to the speed and power required.

Advantages of electric propulsion

The large amount of electric power available for main propulsion can be diverted for cargo or dredge pump operation as well as for bow or stern thrusters and fire pumps of the emergency and support vessel (ESV) described. There is potential for reduction in the size of propulsion machinery spaces, because machinery is smaller and the generators, whether diesel, steam or gas turbine driven, can be located anywhere. One proposal for a liquefied natural gas carrier was for a pair of gas turbine driven generators located at deck level, with electric propulsion motors of 36 000 h.p. (27 000 kW) situated in a very small aft machinery space. Electric propulsion separates the shaft and propeller system from the direct effect of a diesel prime mover and from transmitted torsional vibrations.

Disadvantages of electric propulsion

Higher installation costs and lower efficiency compared with a diesel propulsion system are the likely penalties.

D.c. propulsion motors

The power for direct current motors is limited to about 8 MW so that a.c. machines are used for higher outputs unless resort is made to the installation of d.c. motors in tandem.

Ward-Leonard control

The Ward-Leonard system, as described in Chapter 8, has an a.c. induction motor driving its direct current generator. Field current for both generator and motor is delivered through a three-phase rectifier from the a.c. supply.

The simple Ward-Leonard arrangement for diesel electric propulsion (Figure 11.1) is an all-d.c. scheme with a diesel engine as the prime mover driving the single d.c. generator at constant speed. An exciter mounted on an extension of the generator shaft provides field current both for the generator and for the direct current propulsion motor. The exciter is itself a d.c. shunt generator.

At start-up, the armature windings of the exciter have current generated in them when they pass through the field emanating from the residual magnetism of the exciter poles. The small current generated initially, circulates through the windings of the exciter poles, strengthening their magnetic fields until full output is reached. The current generated in the d.c. exciter is delivered unchanged to its own field poles and to the field poles of the d.c.

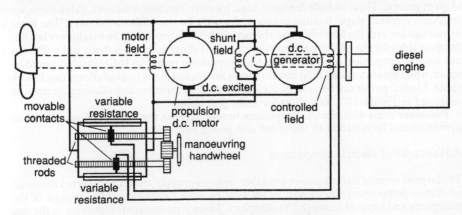

Figure 11.1 *Simple Ward-Leonard system for diesel-electric propulsion*

propulsion motor. It is available to the field poles of the generator, but only through the regulating resistances of the manoeuvring control. If the control contacts are at the mid positions of the resistances, then no current flows to the main generator poles and there is no output from it to the propulsion motor. Rotation of the manoeuvring handwheel and gears turns the threaded bars to move the contacts along the resistances, in opposite directions. As the contacts travel toward the extremities and resistance lessens, current from the exciter flows to the generator field poles. The direction of current flow and the level are used to control the output of the generator and, in turn, the propulsion motor. Propeller speed is proportional to the actual voltage produced in the generator and fed to the propulsion motor.

D.c. constant current system

The generator of the simple Ward-Leonard system described above is dedicated to the propulsion system because control of the propeller is based on variation of the generator field current and voltage output. One advantage of the constant current and other electric propulsion systems is that the generator is involved not with control, but solely with supplying power. The two 610 kW d.c. generators of the constant current system shown (Figure 11.2) supply power at a constant current of 1000 amps for the bow thruster as well as for the two propulsion motors. Other equipment designed for the particular power could also be supplied.

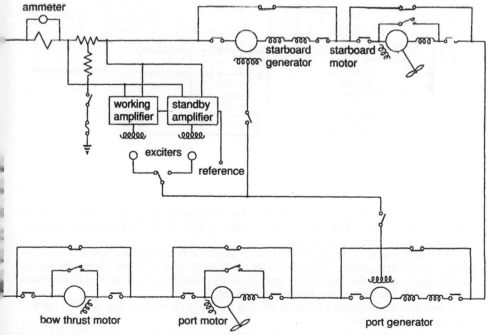

Figure 11.2 *Constant current (diesel) electric propulsion system designed by W. H. Allen Sons and Co. Ltd in 1966*

Excitation for the propulsion generators and motors is provided by either of two motor-driven five-unit exciter sets. Each set consists of a propulsion generator exciter, two propulsion motor exciters and a small high-frequency alternator, all driven by a d.c. motor. The generator exciter can excite one or both generators, simultaneously. Independent control of speed and direction for propulsion motors requires separate excitation, hence the two propulsion motor exciters. The 400 Hz alternator supplies a magnetic amplifier for constant current control.

Constant current systems, as described, were installed in two small heavy-load roll-on/roll-off vessels which were built in 1966 and which at the time of writing are still in service.

D.c. motor supplied from a.c. generators

Direct current propulsion motors installed in more recently built ships are supplied with power from alternators through control and rectification systems. The basic arrangement (Figure 11.3) shows a diesel-driven high voltage a.c. generator with an a.c. exciter on an extension of the shaft.

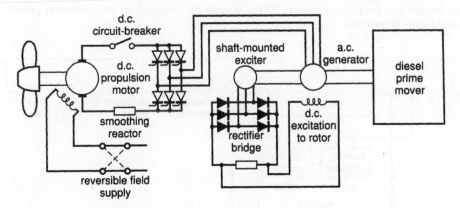

Figure 11.3 *D.c. motor drive from alternator*

Output from the three-phase exciter is rectified and delivered to the alternator rotor as direct current. The level of this current is controlled by an AVR to maintain constant voltage output of the a.c. generator.

The alternating current output from the a.c. generator is delivered to the d.c. propulsion motor armature through a thyristor bridge, as direct current. Control of the gate signals for the silicon-controlled rectifiers (Figure 11.4) alters the level of voltage and hence speed of the motor. The smoothing reactor reduces ripple.

Reversal of the propulsion motor is effected by changing the direction of direct current through the field poles.

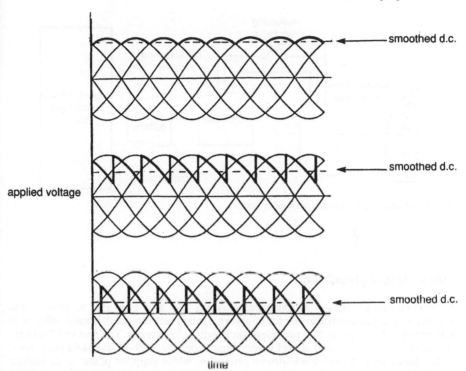

Figure 11.4 *Control of power to propulsion motor*

A.c. propulsion motors

The necessary reversal and speed changes essential for a synchronous motor coupled to a fixed-pitch propeller are obtained in the classic turbo-electric installation (Figure 11.5) by switching two phases of the three-phase power supply to the motor and by altering the speed of the steam turbine, respectively. With this scheme, the variable-speed a.c. generator and the electric propulsion motor provide a system which is a substitute for a gearbox. Manoeuvring is partly by electrical and partly by mechanical means.

The arrangement allows flexibility in the positioning of equipment, but the change of speed and frequency of the turbine-driven alternator is essential to the control of propulsion motor speed. This means that the alternator must be dedicated to the propulsion motor and cannot be used simultaneously to supply power to other motors. The drawback of having the generator involved with control (as with the d.c. Ward-Leonard system) is avoided when constant-speed alternators are used either with constant-speed motors and controllable-pitch propellers, or with propulsion motors with variable speed and direction coupled to fixed-pitch propellers. Control of propulsion motors is provided by modern solid state equipment and generator power can be used for thruster, pumps and other auxiliary machinery.

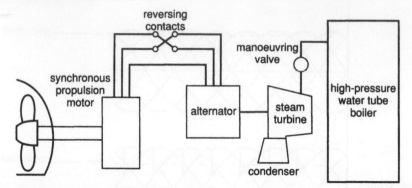

Figure 11.5 *Basic turbo-electric propulsion system*

Turbo-electric propulsion

A turbo-electric propulsion arrangement (Figures 11.5 and 11.6) can be used as the alternative to a reduction gearbox for coupling a steam turbine which is most efficient at high speed and a propeller that provides best results at low speeds. The fleet of T2 tankers built as the tanker equivalent of the Liberty ships during the 1939–45 war had a propulsion system based on a steam-turbine-driven generator, which supplied power to an electric propulsion motor coupled to the propeller. At a time when a rapid tanker-building programme was essential to the war effort, the adoption of an electrical drive overcame the problem of developing a larger reduction gearbox and building up the manufacturing capacity. An advantage of turbo-electric propulsion is that an astern turbine is not required, as reversing is effected through the switchgear.

The component parts of the simple turbo-electric installation are the *propulsion motor* which is coupled to the propeller shaft, the *turbine-driven generator*, and the *boiler* which supplies steam for the turbine. The synchronous generator, intended for high revolutions, has a two-pole-wound rotor and is of relatively small diameter to reduce problems due to centrifugal effect. Generator excitation must be capable of continuous control to provide high excitation and voltage for starting and a varying excitation, as well as voltage for the different speeds and frequencies. The synchronous propulsion motor runs at a slow speed (say 106 r.p.m.) to suit propeller efficiency and has, therefore, a large number of salient poles. Direct current for rotor excitation is provided separately. The propulsion motor has copper bars and end rings in the periphery for starting as an induction motor.

The turbine is warmed through in the normal way and, at the first movement following stand-by, the control valve is opened to admit steam and bring the generator set to idling speed. When the direction contacts are set, the alternator is provided with extra excitation to raise its voltage and the propulsion unit, now temporarily an induction motor, is accelerated to idling speed. The torque of an induction motor at starting varies in proportion to the square of the applied voltage. The propulsion motor now becomes synchronous as the rotor poles are supplied with direct excitation current. The generator excitation is now reduced and kept to an appropriate level as the speed of the turbine and propulsion system is increased as required.

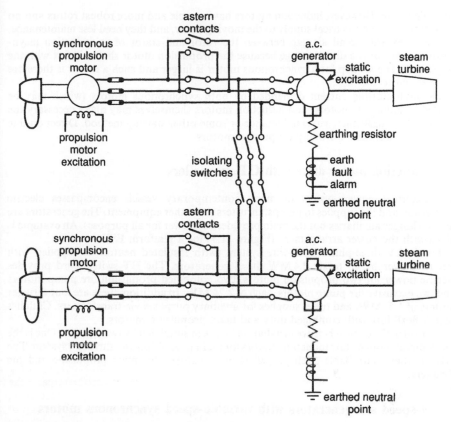

Figure 11.6 *Basic electrical scheme for turbo-electric propulsion*

Prior to stopping/reversing, the system is brought back to idling speed and the excitation is cut off.

The main propulsion alternators (Figure 11.6) have earthed neutrals, with resistance to limit any earth fault current and alarms.

A.c. one-speed drive with C.P. propeller

Controllable-pitch propellers are used to provide speed variation and manoeuvring capability when one-speed synchronous or induction (asynchronous) propulsion motors are installed. The former are more costly than the asynchronous types but also more efficient. Synchronous motors do not diminish the power factor of systems and by increase of their excitation can remedy the decrease of p.f. caused by other equipment.

Because of this non-consumption of reactive power and better efficiency, the generator supplying a synchronous motor does not need to be as large as for an induction motor of equivalent size. However, induction motors have simple and more robust rotors and no requirement for an electrical supply to the moving rotor, and they need less maintenance.

The necessarily small air gap between the rotor and stator of an induction (asynchronous) motor is a disadvantage because the propulsion motor shaft will flex with the ship's hull. The air gap for a synchronous motor is larger and such a motor is therefore more suited to the task.

The high starting current demanded by both synchronous motors (which may be started as induction motors) and induction motors themselves may make necessary the use of auto-transformer, series inductance or some other starting method. Direct on-line starting is acceptable for many propulsion motors.

A.c. induction motor drive with C.P. propellers

The electrical power system of many contemporary vessels encompasses electric propulsion and the supplies to pumps, thrusters and other equipment. The generators are now no longer auxiliaries but the main providers of power for all purposes. An example is given with the power arrangement (Figure 11.7) for a platform ESV.

The 6.6 kV three-phase three-wire system (with insulated neutral) is supplied with electric power by up to six 3.4 MW diesel generators. The HT switchboard provides electric power for two propulsion motors, two combined propulsion/fire pump motors and two exclusive fire pump motors, each of 2.24 MW. In addition, there are four thruster motors of 1.5 MW, and other supplies for auxiliary purposes via transformers. Clutches permit flexibility with combined duty and main propulsion motors.

Control of the comprehensive installation has been simplified by two features. One is the use of direct on-line-start induction motors for main propulsion and thrusters system. The other is that controllable-pitch propellers are employed for main propulsion and the thrusters.

Fixed-speed a.c. generators with variable-speed synchronous motors

The potential for using static electronic equipment to change an electrical supply can be seen from the conversion systems described in Chapter 10 and from the description of an a.c. drive for a d.c. propulsion motor in this chapter. Static frequency converters are used in a number of ship's installations (Figure 11.8) as the controlling intermediary between fixed-speed alternators and variable-speed synchronous propulsion motors. The output from the a.c. generators is delivered at constant voltage and frequency, but for manoeuvring or slow speeds is passed on to propulsion motors at a lower frequency and with voltage adjusted. The speed of a synchronous motor is dictated by the frequency of the current supplied.

Many synchronous drives are based on conversion of the output from fixed-speed a.c. generators first to direct current and then back to a.c. at a lower frequency (the opposite of the converter scheme for variable-speed shaft generators). The vessel, when operating at full speed, will receive power at normal frequency and voltage straight from the switchboard.

Cycloconverter

The cycloconverter method of controlling speed also relies on the ability of the converter to accept current from the switchboard at constant frequency and voltage but to pass this

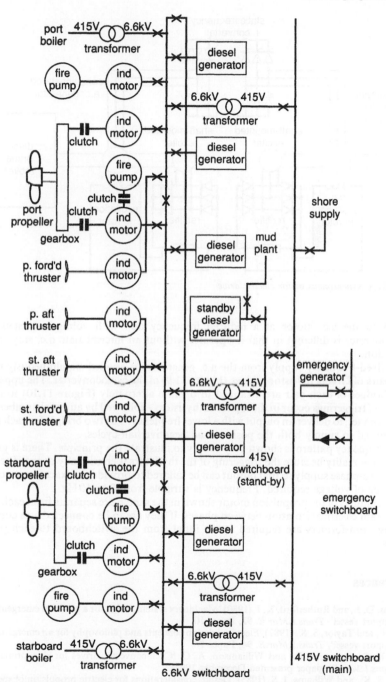

Figure 11.7 *Electrical power arrangement for an offshore emergency and supply vessel*

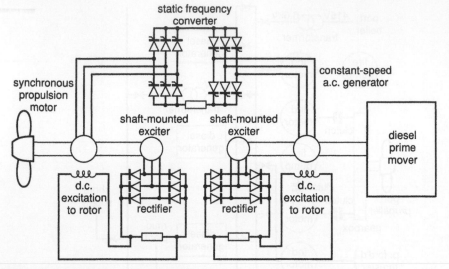

Figure 11.8 *Synchronous motor electric drive*

current to the a.c. motor at a reduced frequency and with voltage adjusted. The cycloconverter is different in that it operates without an intermediate d.c. stage in the conversion.

The fixed-frequency supply from the a.c. generators is applied simultaneously to the three pairs of Graetz thyristor bridges (Figure 11.9) of the cycloconverter. The upper and lower bridges of each pair are arranged to operate alternately (Figure 11.10) so that a number of triggerings occur in the top set of thyristors – followed by an equal number from the bottom set, to deliver an output with a lower frequency. The two bridges for each phase are required to supply both the positive and negative half-cycles.

The frequency pattern is shown very simply to illustrate the principle. There is greater variation in reality because the triggering of the thyristors is continually changed relative to the three-phase supply so that output can be tailored to provide the exact frequency and amplitude of voltage required. Frequency is variable from 0 to 60 Hz.

The windings of the propulsion motor shown in the sketch are separate from each other to maintain electrical isolation between phases. If they are to be connected as a common winding, transformers are required at the input from the switchboard to each pair of bridges.

References

Gibbons, D. J., and Rutherford, K. J. (1980) 'Machinery system design for a platform emergency and support vessel'. *Trans.I.Mar.E.* **92**, Paper 10.

Rush, H., and Taylor, S. K. (1982) 'Electrical design concepts and philosophy for an emergency and support vessel'. *Trans.I.Mar.E.* **94**, Paper 28.

Smith, K. S., Yacamini, R., and Williamson, A. C. 'Cycloconverter drives for ship propulsion'. *Trans.I.Mar.E.* (paper presented December 1992).

Taylor, S. K., and Williams, J. S. (1985) 'Design considerations for electric propulsion of specialist offshore vessels'. *Trans.I.Mar.E.* **97**, Paper 6.

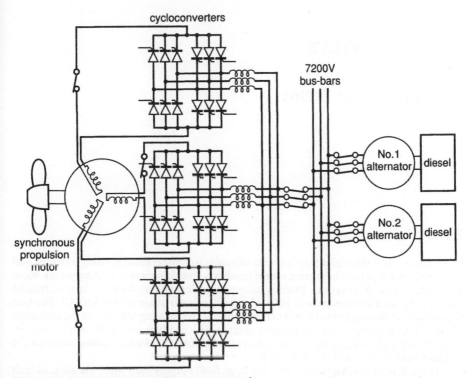

Figure 11.9 *Cycloconverter propulsion motor control*

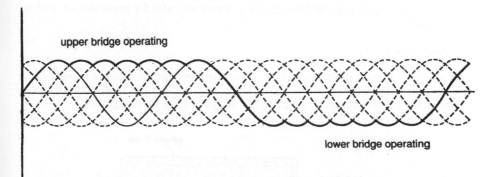

Figure 11.10 *Frequency change using a cycloconverter*

CHAPTER TWELVE

Miscellaneous Items

Fuses

High current flow through a thin fuse wire will raise its temperature, causing it to melt and break the circuit before the current excess reaches a level sufficient to damage other, more substantial parts of the system. Melting temperature depends on the material used (tinned copper in rewireable fuses melts at 1080 °C, the silver in cartridge fuses at 960 °C). The wire is sized so that the normal current is carried without overheating but, due to the resistance of the relatively small wire, that excess current will produce heat sufficient to melt it. **Current rating** gives the normal current that may be carried; **minimum fusing current** is the smallest current that will cause melting.

A fuse will melt much quicker with very large fault current than when the value of fault current is only just above the minimum fusing current. Time/current characteristics are found by testing six or more of the same type of fuse at different currents and plotting the results. The bottom current for the test is not more than 1.05 × minimum fusing current, and the top current is one that will melt the wire in not more than 0.5 of a second. The other test currents are equally spaced between these. Fuses are rated for particular a.c. and/or d.c. voltages.

Cartridge fuses

High rupture capacity (HRC) fuses have silver wire enclosed in a quartz-powder-filled ceramic tube with metal end-caps (Figure 12.1). Arcing when this type of fuse blows is buried in the powder, fusion of which in the arc path helps to extinguish it.

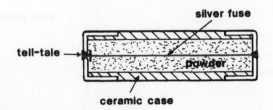

Figure 12.1 *HRC or High Rupture Capacity fuse*

HRC fuses can be used for very high fault levels; deterioration is negligible; they have accurate time/current characteristics and reliability for discrimination; they are safer if accidentally inserted on a fault; there is no issue of smoke or flame; cartridges are sized to ensure that the correct value fuse is fitted.

Semi-enclosed fuses

The rewireable fuse has an insulated carrier for safe handling and containment of the wire in an asbestos-lined tube (Figure 12.2). The wire is easily replaced after operation, but the design is open to abuse as too heavy a wire can be used which could mask a fault and also cause severe arcing if it did operate. Another fault is that of premature failure if the wire is made thinner by oxidation or contact with air, or by being stretched when fitted (a problem with wire made of lead, tin or an alloy of the two).

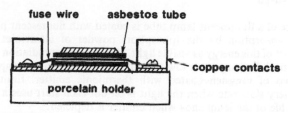

Figure 12.2 *Rewireable fuse*

Fuses in service

Fuses may be used as the only protection in a steady-load circuit, such as for lighting. An a.c. motor with its high starting current and varying load has fuses in each of the supply conductors, but fitted as a back-up for the other forms of protection and to break the circuit in the event of a short-circuit current greater than that which the ordinary contact breakers are designed to interrupt without damage. Very accurate time/current characteristics are needed for fuses used in conjunction with other safety devices, to ensure that the overload trip is allowed time to operate for moderate over-current but that the fuse blows first if there is very high short-circuit current.

Incandescent lamps

High current flow in the coiled high-resistance tungsten filament of an incandescent lamp gives a temperature rise sufficient to produce an almost white glow. Much of the electrical energy supplied, however, is radiated in the form of heat and less than 20% is converted to light. Best efficiency might be obtained with a filament temperature of over 6000 °C; however the best filament material, tungsten, melts at 3400 °C. Above 6200 °C efficiency drops as the wavelength of the radiated energy shortens to the extent that non-visible ultraviolet light is emitted at the expense of visible light.

A near-white-hot filament would quickly burn away in air (as seen when a light bulb breaks), but vacuum bulbs tend to blacken from evaporation of the tungsten. Introduction of the inert gases argon and nitrogen lengthened filament life by reducing evaporation and allowed higher temperature operation and greater efficiency.

Disadvantages of incandescent (filament) lamps compared with the fluorescent type are these. (1) They are more expensive to run, giving less light for the power supplied. (2) They produce about twice the heat for the same light output. (3) Filament glare is a problem in clear bulbs for lighting although it may be an advantage for indicator bulbs. Pearl bulbs are less efficient but give a more even light. (4) Filament lamps have a shorter life and are affected by shock and vibration. (5) Voltage reduction will reduce filament temperature and increase life, but efficiency drops. Higher voltage improves light but shortens life. Rough service lamps are available.

An *advantage* of incandescent lamps is that they do not require a complicated starting circuit and this makes them convenient and cheap for many purposes.

Fluorescent lamps

The inner surface of a fluorescent lamp tube is coated with fluorescent powder. Light is produced from absorption by the (phosphor) powder of ultraviolet radiation and re-emission of part of this energy as visible light. The ultraviolet radiation comes from the effect of an electrical discharge on a low-pressure mixture of mercury vapour and argon, between cathodes of tungsten coated with thermionic emitter. Emitter material is consumed at a very slow rate when the light is on but each start uses a relatively large amount. Useful life of the lamp ends when emitter is depleted.

Switch start circuit

A **glow type starter** is a switch consisting of bimetal contacts separated by a small gap and contained in a gas-filled bulb (Figure 12.3). Circuit voltage is high enough when switched on to produce a glow in the gas and sufficient heat to make the bimetal contacts touch as a

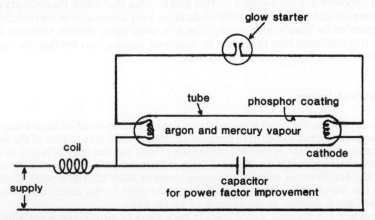

Figure 12.3 *Fluorescent lamp with glow-start circuit*

result of differential expansion. Once the contacts come together, arcing across the gap ceases and the glow is extinguished. The closed contacts now complete a circuit through the inductance coil and cathodes, causing the latter to heat up to red heat. The circuit is broken when the cooling bimetal contacts move apart. This interruption of current in the inductance coil creates a high-voltage pulse that ionises the gas and vapour in the tube and initiates current flow between the cathodes. With current flowing, voltage drops to below the level needed to re-ignite the glow bulb.

A **thermal starter** switch (Figure 12.4) operates on the same principle, in that it opens to produce high voltage by inductance. It also has bimetal contacts but these are closed when cold. When the circuit is energised, current flows through coil, heater and cathodes. Heating caused by the current brings the cathodes to red heat and the switch heater to a temperature high enough to open the bimetal contacts. Interruption of the current in the highly inductive circuit produces the striking voltage to start the lamp.

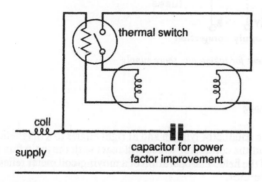

Figure 12.4 *Thermal switch start circuit for fluorescent lamp*

Shore electrical supply

The connection box for taking an electrical supply from ashore is fitted in a position which is easily accessible for the cables and itself connected to the engineroom switchboard by a permanent cable installation. A convenient place may be a locker at deck level that can be reached from either side of the vessel. Figure 12.5 shows equipment for taking an a.c. supply.

Details of the ship's a.c. system, with voltage and frequency, are given in the connection procedure instruction in the terminal box. A ship-to-shore earth is required if the shore power is three-phase with an earthed neutral.

After terminal connections are made, before closing the breaker, phase sequence is checked with the indicator to ensure that motors will not run in the wrong direction. The indicator lamp shows availability of shore power at the switchboard and the supply breaker is closed after the ship's alternators are disconnected. Overload protection is provided by the supply breaker or an isolating switch and fuses. There is usually a lamp to show that the supply is on, together with a voltmeter and ammeter.

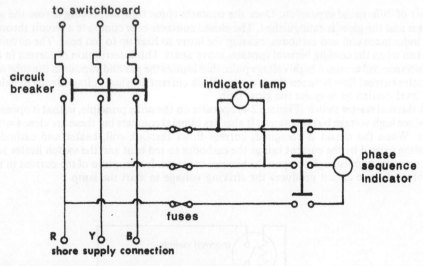

Figure 12.5 *Arrangement for taking a.c. shore supply*

Moving-coil meters

Current supplied to a conductor lying in and at right angles to a magnetic field will set up a magnetic field around the conductor which will react with the main flux and tend to move the conductor out of the field. The operation of a moving-coil meter relies on this principle, as does the electric motor.

The field of a moving-coil meter is provided by a permanent magnet with the flux being strengthened by a soft iron core fixed in the gap by clamps top and bottom. A moving coil, wound on a light frame and mounted on a spindle, is fitted so that the coil sides can rotate in the space between the iron core and the poles. The spindle passes through a hole in the iron core and is supported in bearings at either end. Current from the supply to be tested passes in and out of the coil via hair springs (non-magnetic) which also control the movement. The small angle through which the coil moves is proportional to current flow through the coil and a pointer on the spindle indicates the reading on an even scale. Current in the coil is limited by the small size of the wire to perhaps 69 mA, so that meters intended for use as ammeters measuring current require a low-resistance shunt to bypass the greater part. A resistance in series is required by moving-coil instruments intended for use as voltmeters.

Moving-coil instruments are basically devices for measuring direct current, and their use with alternating current requires that the supply be rectified. Oscillation of the pointer due to current fluctuations or spring vibration is damped by eddy current induced in the light aluminium frame on which the coil is wound.

Moving-coil ammeter

Small current can be measured with a moving-coil instrument without the necessity of a shunt, but otherwise one is needed to bypass the excess current. Shunt size is fixed in meters for switchboards and starter boxes. Test ammeters have several built-in shunts arranged with a switch for changing the meter range, and some have external replaceable shunts.

A shunt must be capable of carrying heavy current without overheating and its resistance must not change appreciably with temperature. False readings are likely if the ammeter is connected to an external shunt with leads of different resistance to those supplied.

Moving-coil voltmeter

Resistance in series with the moving coil is inserted when an instrument is intended for use as a voltmeter. Usually the resistance is mounted inside the instrument. A multi-range meter has a tapped resistance with a selector switch to change the operating range.

The resistance, sometimes called a voltage multiplier or multiplier resistor, prevents large current flow through the instrument, which has only a fine winding on the moving coil. Thus a voltmeter can be connected across the terminals of a 220 volt d.c. generator and the large series resistance will reduce current flow to the level suitable for the meter. Voltage variations will vary the amount of current passing through the resistance and the moving-coil meter will register the changes on a scale marked in volts.

Moving-iron meters

These meters are versatile instruments that can be used for measurement of current or voltage with both d.c. and a.c. equipment.

The repulsion type has a large coil with terminals for connection of the supply to be tested. Current in the coil, whether d.c. or a.c., will set up a magnetic field so that the fixed and moving irons within the field are also magnetised. Both irons are magnetised with the same polarity so that they tend to repel each other. The moving iron, being attached through a lever to the pointer spindle, causes the instrument to register the effect on a scale.

Pointer movement is controlled by hairsprings on the spindle ends, but these meters are often shown in sketches with weights for gravity control. Pointer oscillation is reduced by the air-damping vane.

Electric shock

The effects of severe electric shock and immediate first aid required for its victims are shown by posters on display at high-risk areas such as the switchboard. Resuscitation techniques are also taught in the mandatory first aid courses. Certain conditions increase the dangers from electric shock, and risks are greater when using portable a.c. appliances than with fixed electrical installations.

Current from a steady d.c. source, in passing through the skin, will tend to cause muscular contraction at the initial contact and as contact is broken. Alternating current produces a continuing spasm in the muscles through which current passes, with its change from forward to reverse flow at the rate of 50 or 60 cycles per second. Alternating current has the additional ability to stimulate nerves directly. Most victims of 'serious shock' will have been in contact with a.c. Serious shock results in unconsciousness or worse, requiring resuscitation and medical care.

Alternating current which takes a path through the chest area can, by contraction of the chest and diaphragm muscles, stop the breathing directly, and possibly also indirectly by interfering with the operation of the respiratory control nerves. Similarly, shock in the region of the chest can have direct consequences for the heart, causing stoppage from contraction of the heart muscles. Lesser alternating currents can upset the heart's pumping action by destroying the coordination between the walls of the ventricles (ventricular fibrillation). Current flowing through the body can cause clotting within blood vessels so that tissues are starved of blood. Various nerves may be affected, also the brain or other vital organs could be injured.

Serious shock as a consequence of the above can kill instantly, in so far as stoppage of the heart and breathing are equated with death. However, with the power shut off or the person safely removed from contact, the prompt and continuing application of first aid has a 75% chance of saving life. (With shock, arrest of breathing and heartbeat are not the result of a physical defect but of a temporary condition induced by the electric current, and with only brief contact there may not be serious damage from the current.) Resuscitation to overcome loss of heartbeat and breathing calls for both heart massage and artificial respiration to be employed. An unconscious person who is not breathing must be given artificial respiration. After recovery, victims of shock are kept under close observation because of the likelihood of relapse. Unconsciousness or other forms of distress may be delayed and not follow immediately after a shock which has apparently left the victim only shaken.

Burns

Damage from electrical burns may not appear to be extensive from the surface mark (sometimes just a small whitened area), but the penetration may be deep. Current flow can cause clotting of the blood and destruction of tissue. Most cases of severe burning result from contact with a direct current supply.

Conditions which increase danger to personnel

The involuntary spasm caused by electrical contact on some parts of the body sometimes makes the victim jump away. Contraction in muscles of the hand caused by contact with a.c. can mean there is inability to release the object from which shock is being received, and so contact is prolonged. A current of 12 to 15 mA or more through the muscles is sufficient to make relaxation of the grip impossible and a 10 mA current can be fatal over a long period.

The resistance of dry skin to current flow is fairly high, but that of wet skin much less (the body's internal resistance is very low). Thus in warm conditions danger from shock is greater due to sweat on the skin and this has been a feature of some welding accidents. The resistance of wet skin, if taken as $1000\,\Omega$ per cm^2, would permit current flow from a 220 V supply (1 cm contact area) of $220/1000 = 220$ mA. This is more than enough to be lethal. Obviously a higher voltage would increase the current flow. Other factors which reduce resistance of the skin are poor general health and cuts or other surface damage.

Current flows into the body through the part in contact with a live conductor and then out through another part which is touching earth or another live contact at different potential. The current path may be from one hand to the other, through the chest (resistance between the hands may be $2000\,\Omega$ depending on the area of skin involved), or from hand to foot etc. Current flow into the body is less when the skin is dry; and if there is resistance on the current path between the body and earth this will further reduce or prevent current flow and shock (rubber mats, dry metal-free footwear). There is greater risk when working with electrical equipment in humid or wet conditions; in hot conditions where skin, clothing and even protective leather gauntlets become soaked with perspiration; and when in contact with metal platforms, railings, machinery or a metal workbench.

The effect of electric shock is more serious for someone in poor health with, say, a heart problem.

Shock risk with portable a.c. appliances

Faulty hand-held tools powered by a.c. and having metal casings could impart a lethal shock to the operator where the fault causes the casing to be 'live'. The hand(s) gripping the tool provide a large contact area (possibly damp with perspiration) so that sufficient alternating current might flow to prevent relaxation of the hold, and such a current could result in fatality. The risk of shock is increased if the operator is working in damp conditions and standing on metal plates or touching metal structure. There are similar risks with various types of portable or semi-portable appliances – particularly lead lamps.

The metal casings of portable appliances are connected to earth through the earth wire in the three-core cable and the earth pin in the plug, to give protection against a fault which could make the casing live. Frequently, rough handling of portable equipment not only causes the fault which makes the casing live, but also causes the earth wire to be broken. Thus, when electrical connections and insulation are checked in the course of regular inspection and cleaning, the earth core of the electric cable should also be tested for continuity (i.e. with one terminal of the tester on the metal casing of the appliance and the other on the earth pin of the plug).

Shock risk from portable tools is greatly reduced if the power supply is taken from the secondary winding of a transformer used to step down the mains supply to a suitable lower voltage, with the mid-point of the secondary winding earthed. If the secondary voltage is limited to 110 V for operation of the single-phase appliance, then the potential shock voltage between the casing and earth is limited to 55 V. (Secondary voltage can be made lower if required.) Double-pole switches are fitted to control single-phase appliances fed in this way.

Flexible cable for portable tools and equipment is reinforced and given extra protection by a rubber tube where it enters the appliance. Here and at the plug end the cable is subject to bending and pulling: it can also be damaged along its length by being pinched or cut by sharp edges and by touching a hot surface or lying in oil, chemical or water. Sometimes the cable is cut by the the tool being used. Damage to the cable can cause shock in a number of ways, or an earth or a short-circuit.

During cable inspections for damage to the insulation and continuity (particularly in the earth wire), other checks are also made. Live and neutral wires must be correctly fitted to terminal points so that the switch is on the live side and the appliance is isolated from the power supply when switched off. Switch operation is tested. Fuses should be exposed to ensure they are of the right size.

Extension leads must be arranged so that, when plugged into the power supply, the free end has a socket for the three-pin plug of the power tool. If fitted wrongly with a plug, the free end of the extension lead would be highly dangerous.

A small shock from a portable tool can sometimes cause a fall and injury (e.g. to someone working on a ladder).

Merchant Shipping Notice M752 (Safety – Electric Shock Hazard in the Use of Electric Arc Welding Plant) points out the risks involved with the use of arc-welding equipment in hot, damp conditions within a restricted space surrounded by the earthed steel structure of the ship. Three fatalities are cited, each resulting from use of a.c. sets with an open voltage of 70 V or so. The Notice states that d.c. welding sets of the same voltage are safer (unless derived from a.c. and having unsmoothed ripple). Availability of voltage reduction safety devices for installation with a.c. welding plant to make it safer is stressed. These safety devices limit the open-circuit voltage to 25 V until electrode contact is made to strike the arc and then full open-circuit voltage is turned on. The Notice also recommends the use of fully insulated electrode holders, adequate protective clothing including non-conducting rubber boots, good lighting and the display of first aid instructions on a wall chart. Operational advice is given, in particular with regard to handling electrodes, which should not be inserted into a live holder.

Index

Printed and bound by CPI Group (UK) Ltd, Croydon, CR0 4YY

03/10/2024

01040432-0006